BANGLADESH'S AGRICULTURE, NATURAL RESOURCES, AND RURAL DEVELOPMENT SECTOR ASSESSMENT AND STRATEGY

MARCH 2023

ADB

ASIAN DEVELOPMENT BANK

© 2023 Asian Development Bank
6 ADB Avenue, Mandaluyong City, 1550 Metro Manila, Philippines
Tel +63 2 8632 4444; Fax +63 2 8636 2444
www.adb.org

Some rights reserved. Published in 2023.

ISBN 978-92-9270-050-8 (print); 978-92-9270-051-5 (electronic); 978-92-9270-052-2 (ebook)
Publication Stock No. TCS230050
DOI: http://dx.doi.org/10.22617/TCS230050

The views expressed in this publication are those of the authors and do not necessarily reflect the views and policies of the Asian Development Bank (ADB) or its Board of Governors or the governments they represent.

ADB does not guarantee the accuracy of the data included in this publication and accepts no responsibility for any consequence of their use. The mention of specific companies or products of manufacturers does not imply that they are endorsed or recommended by ADB in preference to others of a similar nature that are not mentioned.

By making any designation of or reference to a particular territory or geographic area, or by using the term "country" in this publication, ADB does not intend to make any judgments as to the legal or other status of any territory or area.

Corrigenda to ADB publications may be found at http://www.adb.org/publications/corrigenda.

Notes:
In this publication, "$" refers to United States dollars.
ADB recognizes "Korea" as the Republic of Korea.
The fiscal year (FY) of the Government of Bangladesh ends on 30 June. "FY" before a calendar year denotes the year in which the fiscal year ends, e.g., FY2022 ends on 30 June 2022.

An agricultural system that increases productivity and value addition and is sustainable and climate resilient will contribute to poverty reduction, food security, and continued structural transformation of the economy. (All photos are from ADB).

Cover design by Maro de Guzman.

Contents

Figures and Boxes

Acknowledgments

This country sector assessment was prepared by a team comprising South Asia Environment and Natural Resources (SAER) Division staff Andrew Brubaker, Bakhodir Mirzaev, and Mafalda Pinto. Consultant Arun Saha in Bangladesh supported the team in preparing the document. Consultant Mahfuzuddin Ahmed provided strategic and technical advice on the draft assessment. South Asia Department (SARD) colleagues in other sector divisions and the Private Sector Operations Department (PSOD) provided important background and contextual information for better assessing and identifying crosscutting agriculture and natural resources investment opportunities supporting an integrated Asian Development Bank (ADB) approach. SAER staff Olivier Drieu, Sunae Kim, Sanath Ranawana, and Lance Gore and colleagues from the Bangladesh Resident Mission (BRM) Soon Chan Hong, Pushkar Srivastava, Amrita Kumar Das, and Nasheeba Selim provided valuable inputs and background information at the early stages of the assessment and commented on the final draft assessment. The Sustainable Development and Climate Change Department's Rural Development and Food Security, Climate Change, and Water Thematic Groups made important contributions and insights as reviewers. The team worked under the guidance of Director Mio Oka, who initiated the assessment.

The team would like to give special thanks to BRM and PSOD staff, especially Mohammed Sayeed Haque, Bidyut Saha, and Edwin Enriquez David, who provided data and other information on ADB's sovereign and nonsovereign portfolio, and to SARD Climate Change Decision Support System team (Denver G. Vender) and the Manila Observatory (Celine Vicente) for the preparation of maps on the estimate of crops affected by droughts and floods in Bangladesh.

The team also acknowledges the support from government officials, development partners, research institutes, and other country representatives who made themselves available for consultations held virtually due to travel restrictions associated with coronavirus disease.[a]

---

[a] Government institutions: The Ministry of Local Government, Rural Development and Co-operatives; Local Government Engineering Department; Ministry of Water Resources; Bangladesh Water Development Board; Ministry of Agriculture; Department of Agriculture Extension; Department of Agricultural Marketing; and Ministry of Chittagong Hill Tracts Affairs as well as financial intermediate institutions such as Palli Karma Sahayak Foundation. Development partners: Asian Infrastructure Investment Bank, Food and Agriculture Organization of the United Nations, International Fund for Agricultural Development, Islamic Development Bank, Japan International Cooperation Agency, Kreditanstalt für Wiederaufbau (Credit Institute for Reconstruction), The Embassy of the Kingdom of the Netherlands Water Program in Bangladesh; United States Agency for International Development; and World Bank. Research institutes, International Crops Research Institute for the Semi-Arid Tropics and International Water Management Institute.

Abbreviations

ADB	Asian Development Bank
ANR	agriculture, natural resources, and rural development
AFSRD	agriculture, food security, and rural development
BDP2100	Bangladesh Delta Plan 2100
COVID-19	coronavirus disease
CPS	country partnership strategy
CSA	country sector assessment
DRM	disaster risk management
DTW	deep tube well
FY	fiscal year
GDP	gross domestic product
HVC	high-value crop
ICT	information and communication technology
LCG	local consultative group
NBS	nature-based solutions
O&M	operation and maintenance
PRAN	Programme for Rural Advancement Nationally
R&D	research and development
SDG	Sustainable Development Goal
SMEs	small and medium-sized enterprises
STW	shallow tube well

Executive Summary

This country sector assessment (CSA) considers the context and issues facing the agriculture, natural resources, and rural development (ANR) sector in Bangladesh; it also identifies needs and opportunities for Asian Development Bank (ADB) support. Our findings inform the bank's ANR strategy for Bangladesh. This CSA reviews (i) sector performance, including constraints and opportunities; (ii) government policies, strategies, and plans for the sector; and (iii) ADB and other development partners' support and experience in the sector. Based on our analysis and findings, this assessment identifies strategic directions and investment priorities for ADB in Bangladesh's ANR sector, in line with the country partnership strategy (CPS) 2021–2025.

The Agriculture, Natural Resources, and Rural Development Sector and the Economy

Bangladesh's economy has grown substantially in the last 20 years, dramatically changing its structure and the role of agriculture in promoting continued growth, food security, and poverty reduction. The proportion of gross domestic product (GDP) generated by agriculture fell from about 18% to 13% between 2009–2010 and 2018–2019, as contributions from manufacturing and services sectors increased. This growth, diversification of the economy, and urbanization have changed food demands and expectations from the ANR sector. Responding to this challenge becomes more difficult as the overall performance of agriculture has stagnated or regressed because of slow growth in the crop subsector—mainly due to seasonal climate shocks, low productivity, and profitability.

Despite its declining contribution to national GDP, agriculture still provides more than 40% of total employment in Bangladesh, and it plays a vital role in promoting growth of manufacturing and services. An estimated 85% of the poor population lives in rural areas. Agricultural development is crucial for improving livelihoods of the rural poor—generating employment, increasing household income, and reducing poverty. However, agriculture and rural communities remain vulnerable. The country is increasingly exposed to climatic shocks, and agriculture is expected to be one of the most affected sectors due to more flooding, droughts, and saline intrusion worsened by rising sea levels. These changes threaten not only future agricultural production and growth, but also the quality, integrity, and sustainability of natural resources. The sector has been further affected by the coronavirus disease (COVID-19) pandemic and food-price inflation, which threatens to reverse gains in poverty reduction and food security made in recent decades.

Collectively, these challenges have renewed Bangladesh's focus on revitalizing and improving the resilience of agriculture. Significant potential exists for agricultural development and diversification. However, to address sector challenges, a concerted effort and investment are needed; structural transformation could bring a return to higher growth. Increasing productivity and agricultural commercialization could also facilitate broad-based growth and poverty reduction. Building a stronger and more resilient ANR sector is also vital for adaptation to the impacts of climate change. As such, it is increasingly urgent to integrate climate solutions throughout the sector, and address persistent problems: low productivity and sustainability, high postharvest losses, limited value addition, and food safety concerns.

State of Agriculture

Bangladesh's agriculture has focused on achieving self-sufficiency, and is dominated by the production of rice, largely by smallholder farmers. Production is slowly moving towards greater diversification with high-value crops such as fruits and vegetables, livestock, and fisheries, as demand has increased. However, the overall share of production remains small, relative to rice. Irrigation has been important for expanded rice production, but further modernization and efficiency improvements are needed to cope with demand, overcome climatic impacts, and allow more crop diversification.

Education, research, and extension—as well as other facilitators, such as financial investors—are focused on supporting rice production. Efforts are required to direct these institutions and services towards higher-value commodities. In support of such value addition, market structures need to provide adequate frameworks and systems for grading, storage, and food safety. While support to traditional public sector institutions will have to set policy and build the enabling environment, information and communications technology (ICT) with digitalization can help to overcome traditional constraints (e.g., market and weather information). Engaging private sector enterprises and supporting pro-poor value chains are necessary to accelerate diversification and value addition.

Bangladesh's Unique Geography and Natural Resources Determine Its Agriculture

Bangladesh benefits from—and is challenged by—one of the world's most extensive river systems, with a large delta. Due to flat topography, the rivers and delta are dynamic—influenced by earthquakes, flooding, human activity, and increasingly by intense weather events and climate change. Earthquakes upstream have impacted the river by releasing vast amounts of sediment in the river. During the monsoon, 55%–60% of the country is inundated, while in the dry season water is very scarce. Bangladesh's coastal areas are also prone to cyclones, storm surges, and sea level rises.

Water resources management is important for successful agricultural production, and to support vulnerable rural communities. Flood control, drainage, supply of irrigation water, and integrated water management are the main components of water resources management. In addition, disaster risk management, conservation, and closer cooperation with Bangladesh's riparian neighbors (India, Nepal, Bhutan, Myanmar, and the People's Republic of China) are strategic priorities. Over the last 3 decades, agriculture has benefited from significant investments in large flood control, drainage, and irrigation schemes, utilizing mostly surface water for irrigation. Unfortunately, the management and sustainability of these investments has been poor. Further, the scale of investments must increase to meet agricultural demands and climate challenges.

Country Plans

The Government of Bangladesh has developed several strategic plans and policies identifying ambitious goals for future decades. All strategies recognize the importance of modernizing the agriculture sector, developing further resilience to climate hazards, and managing natural resources sustainably. The *Perspective Plan of Bangladesh 2021–2041* aims to eradicate poverty and attain high-income status for the country by 2041. The plan's premise is that as Bangladesh rapidly transforms its economy and society, conscious management of key natural resources—land, water, forestry, natural habitats, and air—is crucial to avoid their depletion and degradation. It also recognizes that modernizing agriculture through diversification, including through the expansion of the fisheries and livestock subsectors, and improved climate resilience are essential to achieve food security and nutritional balance.

The strategic focus of agriculture in the *Bangladesh Delta Plan 2100* and the *Eighth Five-Year Plan (2021–2025)* is on accelerating the commercialization of agriculture to ensure the sustainable supply of safe, diversified food products. The emphasis is on productivity gains, diversification, high value addition, agribusiness and entrepreneurship development, and efficient and sustainable natural resources management with climate-resilient rural infrastructure development. These strategic plans promote the sustainable intensification and diversification of climate-resilient agricultural production systems that are more integrated with local and global markets as they aim to ensure food and nutritional security, while boosting rural livelihoods for people, including women and vulnerable communities. They emphasize the need for improving irrigation efficiency and water productivity to ensure optimal use of available water resources. This strategy for water resources management is based on (i) rational management and optimal use of water resources to reduce vulnerability and ensure the availability of water for economic purposes, health, and hygiene; (ii) conservation of aquatic and water-dependent ecosystems; (iii) institutional reform of water sector agencies to achieve efficient, integrated water management with sustainable operation and maintenance, and (iv) inclusive stakeholder participation.

ADB Priorities

ADB's ANR support in Bangladesh will respond to the evolving context and changes in the economy, building on its experience. This approach recognizes the ongoing structural transformation of the economy, changing roles of the ANR sector, and existential threats posed by climate change. The sector strategy, while aligning with the CPS, will aim to reinforce ADB's comparative advantages in supporting water resources management and rural infrastructure development, making better use of the knowledge and skills in areas such as private sector development, finance, governance, and education to support productivity improvements, value addition, and commercialization of agriculture. It will also aim to expand support for ongoing and emerging needs identified in the previous CPS related to climate resilience and adaptation, as well as other crosscutting issues such as gender.

ADB's ANR strategy in Bangladesh will focus on promoting rural development, ensuring food security, and improving rural livelihoods by boosting the sector's productivity and competitiveness resulting in greater agricultural commercialization and increasing value addition. This strategy is underpinned by four pillars: (i) agricultural commercialization and value chains, (ii) market connectivity, (iii) water and natural resources management, and (iv) crosscutting One ADB priorities.[b] Environment and climate issues are integrated

---

[b] The One ADB approach promotes greater integration of its operations across sectors for achieving better development outcomes in line with Strategy 2030.

throughout the program, given the importance of natural resources for agriculture. This strategy, through the One ADB approach, will also address crosscutting concerns such as the policy and regulatory enabling environment for agribusiness, institutional and technical capacity, and gender. In addition, ADB's ANR strategy provides an entry point for ADB and the Government of Bangladesh to develop the sector's pipeline and identify opportunities for collaboration, strengthening synergies with government and development partner programs.

In supporting agricultural commercialization and value chains, the main objective will be to increase smallholder productivity and competitiveness from subsistence or semi-subsistence levels to producing commercial agricultural products and meeting value chain requirements. ADB will support smallholder farmers to (i) increase productivity and improve sustainability through access to inputs that are better adapted to evolving climatic and market conditions, (ii) meet market quality and volume requirements for or high-value crops and those with potential for processing and value addition, and (iii) better access to integrate with value chains by addressing key weaknesses and constraints to participation. For example, to address the challenges of aggregation and economies of scale, efforts will be made to strengthen connections between farmers and intermediaries (e.g., cooperatives) with value chain participants and access to finance. ADB may undertake strategic diagnostic assessments to identify specific commodities and sources in the enabling environment for support. For commodities, this would identify key opportunities and constraints, as well as major public and private sector actors, which may be supported or engaged as strategic partners. In the enabling environment, diagnostics may consider policy, regulatory, or institutional barriers (e.g., related to food safety) inhibiting development of the sector or specific value chains. A key feature of these assessments will be consideration of climate-related risks to key activities in the value chain.

For market connectivity and access, the focus will be on supporting market infrastructure for agricultural products and the development of value chains, enhancing transportation infrastructure to improve farmers' access to inputs, market, and value chain linkages. This will include rehabilitating, upgrading, climate-proofing, and improving the maintenance of rural roads. It will also consider inland water transport, given the country's riparian nature and potential efficiencies of such water transport. In addition, ADB will support investments that improve rural connectivity. Market infrastructure and logistics services supporting agribusiness development will be nurtured to reduce postharvest losses, while investments in market, storage, and processing infrastructure will be encouraged. ADB will also support institutional arrangements and capacity for managing and maintaining such infrastructure, particularly where both public and private sectors can be engaged.

Support for strategic water, land, and natural resources management remains an ADB priority, given the importance of natural resources for agriculture, and Bangladesh's vulnerability to climate change impacts. Irrigation and water resources management investments will fund adaptation to climate change, modernization of irrigation systems for better use of water, and sustainable management of water resources and infrastructure. While addressing these challenges, interventions will be more inclusive, building community participation, improving irrigation service delivery, and increasing accountability to farmer organizations. Institutional capacity strengthening for strategic operation and maintenance planning and implementation will remain a key priority. In response to threats posed by climate change, a more holistic approach and understanding of water and natural resources management will be pursued. Increased emphasis on a climate-resilient approach will also be extended to upgrading support for flood control, riverbank protection and stabilization, and adaptive basin management. The mitigation of flood, coastal, and riverbank erosion will continue to be a priority, complemented by new methods of natural resources management and nature-based solutions: the focus will be on improving disaster risk management (DRM) with better land use, planning, and early warning systems. Opportunities to use more nature-based solutions and green infrastructure to complement traditional frameworks will be explored. ADB will also integrate environmental restoration activities—such as reforestation and wetlands and coastal repair—to align with sustainable livelihood approaches.

Recognizing agriculture's multifaceted nature and complexity, ADB support will also address crosscutting issues and link with the bank's other sector and thematic interventions. Key sectors identified for potential integrated ADB support include (i) energy (e.g., solar-pumped groundwater), (ii) education (e.g., tertiary education), (iii) finance (e.g., lending to rural small and medium-sized enterprises), (iv) public sector management (e.g., safety nets and food security), (v) transport (e.g., reduced commodity transport costs via rail and inland waterways), and (vi) urban (e.g., integrated watershed management and urban–rural linkages). ANR sector interventions will also aim to improve (i) governance and capacity development, (ii) gender equity and mainstreaming, (iii) maternal health and nutrition, (iv) knowledge solutions, (v) innovative research and development of climate-resilient approaches and production technologies, and (vi) partnerships, such as between participants in agricultural value chains, to enhance the productivity and efficiency of the sector.

I. Sector Assessment: Context and Strategic Issues

A. Introduction

This country sector assessment (CSA) comes at a critical time for Asian Development Bank (ADB) support to Bangladesh for agriculture, natural resources, and rural development (ANR). The country's economic growth has accelerated substantially in the last 20 years, dramatically changing the structure of the economy and the role of agriculture in promoting continued growth, food security, and poverty reduction. This growth and urbanization have changed food demands, increasing the urgency of addressing persistently high postharvest losses and food safety concerns. Moreover, Bangladesh is increasingly vulnerable to climatic shocks and changes that threaten not only future agricultural production and growth, but also the quality, integrity, and sustainability of natural resources. The sector has been further affected by the coronavirus disease (COVID-19) pandemic, which threatens to reverse gains in poverty reduction and food security. ADB's Strategy 2030 recognizes the proliferation of challenges for the region, including climate change, and the need for more targeted, integrated support.[1]

This CSA aims to assess the context and issues facing the ANR sector in Bangladesh, better understand the country's needs, and identify opportunities for ADB support. It considers sector performance, government policies, current priority areas and strategy—as well as plans for ANR, constraints and opportunities, with the support and experience of ADB and other development partners. More specifically, this assessment examines ADB's contributions to the country's agricultural development and aims to guide the bank in identifying and prioritizing future investments in the sector in line with the country partnership strategy (CPS) 2021–2025.[2] Based on our analysis and findings, this CSA highlights strategic directions and investment priorities for ADB in Bangladesh's ANR sector.

B. Overall Context

Overview

Bangladesh is part of the South Asian subcontinent. It is located along the eastern side of the Bay of Bengal and borders India to the north and east, and Myanmar to the southeast. It has an area of 147,570 square kilometers, and a population of 167.6 million (as of 2020), 61.8% of which live in rural areas.[3] It is the most populous and the largest of the 10 most densely populated countries in the world. The country is located within the floodplains of the rivers Ganges, Brahmaputra, and Meghna and is vulnerable to natural hazards such as seasonal flooding,

---

[1]   ADB. 2018. *Strategy 2030: Achieving a Prosperous, Inclusive, Resilient, and Sustainable Asia and the Pacific*. Manila.

[2]   ADB. 2021. *Country Partnership Strategy: Bangladesh, 2021–2025—Sustain Growth, Build Resilience, and Foster Inclusion*. Manila.

[3]   World Bank. World Bank Data (accessed 2 December 2021). 2020 value as per World Bank Data.

droughts, and cyclones. The Global Climate Risk Index ranked Bangladesh as the seventh most vulnerable to natural hazards between 1999 and 2018. Despite these challenges, Bangladesh possesses a year-round favorable environment for agriculture with fertile soils, flat topography, and ample water resources.

Bangladesh substantially reduced poverty from 48.9% in 2000 to 20.5% in 2019 (Figure 1). This was achieved by raising per capita income, increasing agricultural production, improving quality of education and health services provision, and information and communication technology (ICT) development. These advances elevated Bangladesh to lower middle-income country status in 2015.[4] Bangladesh aims to continue this growth trend to achieve upper middle-income country status by 2031 and high-income country status by 2041. Nevertheless, the poverty headcount in rural areas (17.4% of the rural population) is more than double the headcount observed in urban areas (6.2% of the urban population).[5]

Importance of the agriculture sector for the economy.[6] The substantial gross domestic product (GDP) growth in recent years (Figure 1) was mainly driven by the manufacturing and services sectors. The contribution of agriculture to GDP has declined steadily since Bangladesh's independence in 1971, from 58.4% in fiscal year (FY) 1973–1974 to 12.6% of total GDP in FY2020 (footnote 5). Despite the decline of agriculture's contribution to GDP in the last 2 decades (Figure 1), the ANR sector remains critical. With increasing agricultural production,

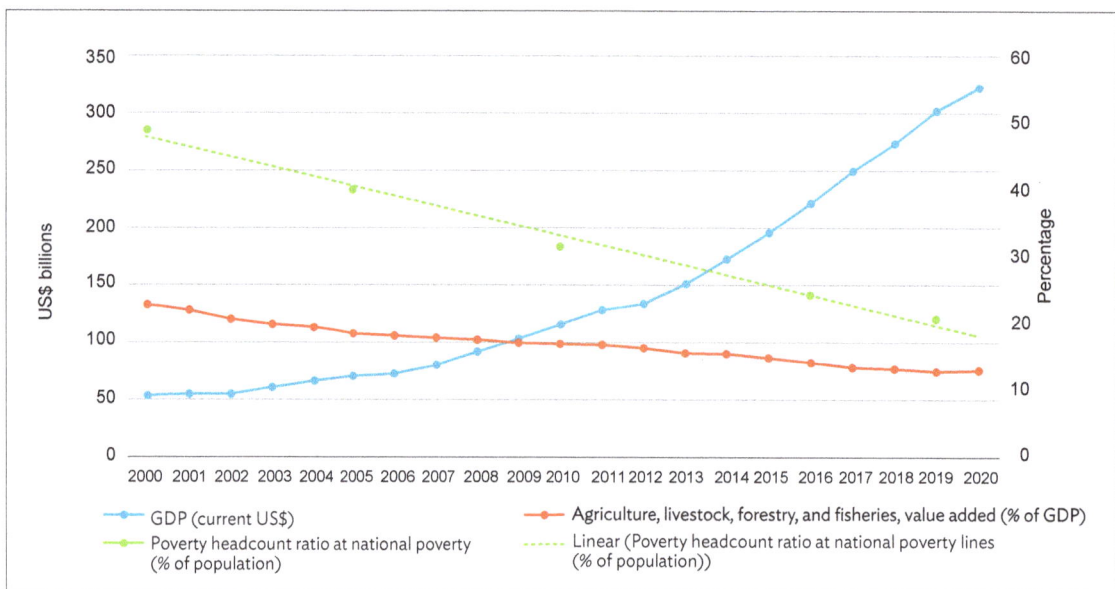

Figure 1: Bangladesh's Gross Domestic Product, Agriculture, and Poverty, 2000–2020

GDP = gross domestic product, US = United States.

Sources: 2020 value for poverty headcount ratio at national poverty is from Asian Development Bank, 2021; other values World Bank Data Bank, World Development Indicators, 2021. (accessed 21 April 2021).

---

[4]   World Bank. 2021. The World Bank classifies countries by gross national income for operational purposes. Middle-income countries are those with $1,026 to $12,535 in per capita gross national income.

[5]   World Bank. World Bank Data (accessed 19 April 2021). Rural and urban poverty headcount ratio is given at $1.90 a day (2011 purchasing power parity).

[6]   The agriculture sector in Bangladesh comprises crops, livestock, fisheries, and forestry.

the country is largely self-sufficient in staple food production (rice). The sector also employs approximately 40% of the labor force and makes significant contributions to poverty reduction and food security through its forward and backward linkages to the rest of the economy.[7] This gives ANR an economic impact much greater than suggested by only considering agriculture's share of GDP (footnote 2).

Structural transformation. Over the last 3 decades, Bangladesh has experienced a moderate but steady structural transformation between and within sectors. The economy is moving from labor-intensive industries such as traditional agriculture to capital-intensive industries like manufacturing (e.g., textiles) and services (e.g., rail and air transportation, water supply and sanitation). In the ANR sector, this structural transformation is marked by the steady progression from traditional subsistence agriculture to commercially viable activities to meet demand for diversified foods from the growing middle class. Recently, rural farmers have started to engage in off-farm activities and employment to supplement on-farm income due to the uncertainty of investment in on-farm activities. Modernization of on-farm production technologies including mechanization related to crops, livestock, and fisheries; expansion of ICT networks; connectivity; and access to electricity and other basic services in rural areas work as drivers of growth and change in the rural economy. In general, the transition to economically viable commercialized farming is progressing well. However, an apparent mismatch between skill or education and demands of the job market suggests a need to provide greater emphasis on skill-oriented training programs, particularly for youth.[8]

Climate change. Due to its low-lying topography, deltaic nature, and location at the base of the Himalayas on the Bay of Bengal, Bangladesh is extremely vulnerable to natural hazards. The impacts in frequency and severity of these hazards are expected to increase due to climate change, with estimates indicating that Bangladesh will lose 2.0% of its annual GDP due to climate change by 2050. Agriculture is expected to be one of the most affected sectors due to increased levels of flooding, droughts, and saline intrusion which is worsened by rising sea levels. In addition, the incidence of insect pests and diseases is also expected to increase due to higher temperatures, humidity, and radiation. Under moderate climate scenarios, crop production will decline by 17% for rice and 61% for wheat.[9] This will largely be due to increased stress from more frequent extreme heat days and periods of drought. Inland and coastal fisheries will similarly be affected by periods of drought and more intense monsoons and cyclones, resulting in flooding and increased saltwater intrusion. Additionally, the predicted increase in temperature and humidity is expected to reduce labor efficiency. These effects are being exacerbated due to unplanned urbanization and development and will negatively influence the food security, livelihoods, and lives of millions of people in the country.[10]

Governance. Governance, fragmented policies, and institutional coordination challenges are prevalent to various degrees in Bangladesh. Institutional silos and lack of accountability in the government's administration system are underlying concerns. Complex bureaucratic systems contribute to governance challenges to a great extent. Institutional reforms to establish strict accountability systems and compliance at all levels are considered highly important. According to Transparency International's Corruption Perceptions Index, Bangladesh ranks 146th out of 180 countries in terms of corruption prevalence, with a ranking of 1 the least and 180 most corrupt. With a Corruption Perceptions Index score of 26 out of 100, the country is well below the Asia and Pacific regional average of 45.[11,12]

---

[7]   Government of Bangladesh, Bureau of Statistics. 2018. *Labour Force Survey 2016–17*. Dhaka.

[8]   International Labour Organization. Bangladesh Statistics (accessed 21 April 2021).

[9]   Estimates considering only changes in temperature, moisture regime, and carbon dioxide fertilization.

[10] M. Ahmed and S. Suphachalasai. 2014. *Assessing the Costs of Climate Change and Adaptation in South Asia*. Manila: ADB.

[11] The results are given on a scale of 0 (highly corrupt) to 100 (very clean).

[12] Transparency International (accessed 19 April 2021).

Coronavirus disease pandemic. The COVID-19 pandemic has had a severe impact on Bangladesh's economy and its population's livelihoods. The Eighth Five-Year Plan (FY2021–2025) estimates substantial losses in GDP ($9.34 billion), exports ($8 billion), investment ($5.84 billion), and tax revenues ($2.34 billion) due to increased unemployment and significantly lowered economic activities.[13] The restricted economic activities contributed to reversing the gains in reducing poverty and food insecurity, while the pandemic also increased the importance of agriculture as a main source of income for households, rising from 26% in 2018 to 29.4% in 2020.[14] COVID-19 effects were muted due to remittances, family farms as a safety net for urban migrants, and the government's various stimulus packages.[15] GDP growth in 2020 fell to 3.4%; however, it recovered to 6.9% in FY2021 and the recovery is forecast to continue for 2022.[16]

Lockdown and other measures to combat the COVID-19 pandemic have significantly affected agricultural supply chains and food consumption behavior. The crisis increased temporary unemployment; seriously impeded farmers' ability to sell products, procure raw materials, and hire labor; and hampered under-developed productive capacities, particularly for poultry and dairy. Various government incentive packages, including low-interest agricultural credit for farmers and entrepreneurs with small and medium-sized enterprises (SMEs), are assisting recovery. Many general workers and skilled professionals have returned to rural areas and started or reengaged in on-farm commercial agricultural activities. A further effect is global inflationary pressure on commodities and increased food prices, with urban areas being more affected by price volatility than rural areas.[17]

State of Agriculture

Agriculture in Bangladesh is dominated by the production of rice, which is carried out largely by smallholder farmers.[18] Rice is cultivated throughout the year with different varieties (Boro, Aus, and Aman) most appropriate for each of the three growing seasons. Rice is planted in about 80% of the cultivated areas suitable for crop production in the agroecological context. However, production is moving toward greater diversification and high-value crops (HVCs). The overall production of HVCs (such as fruits and vegetables, which are mostly grown in flood-free areas), livestock, and fisheries increased significantly over the last decade due to high market demand but remains small relative to the share of paddy to overall production (Figure 2).

Food price increases. Bangladesh experienced a sudden and rapid surge in inflation in 2021 that continued into 2022. Food prices are approaching their highest levels in the last decade and are expected to remain high, due more to wheat than rice, and also due to inflation for nonfood items.[19] Price inflation of agricultural foods and commodities, along with agricultural inputs, hampered farmers' purchasing capacity for food, contributing to their increased food insecurity. The COVID-19 pandemic had a significant impact on the economy, shutting factories and limiting economic activities. However, given the importance of agriculture and its role in most livelihoods, agriculture remained uninterrupted, and food and nutrition security were maintained despite the pandemic. The sector is largely dependent on fuel and fertilizer imports, which have increased dramatically in price; Bangladesh's

---

[13]   Government of Bangladesh, Planning Commission. 2020. *Eighth Five Year Plan*, FY2021–FY2025. Dhaka.

[14]   S. Raihan et al. 2021. *COVID-19 Fallout on Poverty and Livelihoods in Bangladesh: Findings from SANEM's Nationwide Household Survey (November–December 2020)*. Dhaka: South Asian Network on Economic Modeling.

[15]   ADB. 2020. *COVID-19 Policy Database*. Manila. The stimulus packages are estimated to total $13.97 billion (4.3% of GDP).

[16]   ADB. 2022. *Asian Development Outlook (ADO) 2022: Mobilizing Taxes for Development*. Manila.

[17]   M. Chowdhury et al. 2021. The Impact of COVID-19 Pandemic on the Inflation Dynamics of Bangladesh: Lessons for Future Economic Policy Formulation. *Bangladesh Bank Policy Note Series*. Dhaka: Bureau of Statistics.

[18]   About 80% of farmers in Bangladesh cultivate less than 1 hectare of land.

[19]   Government of Bangladesh, Food Planning and Monitoring Unit, Ministry of Food. Ensuring food security for all people at all times.

Figure 2: Crop and Livestock Production, 1961–2019

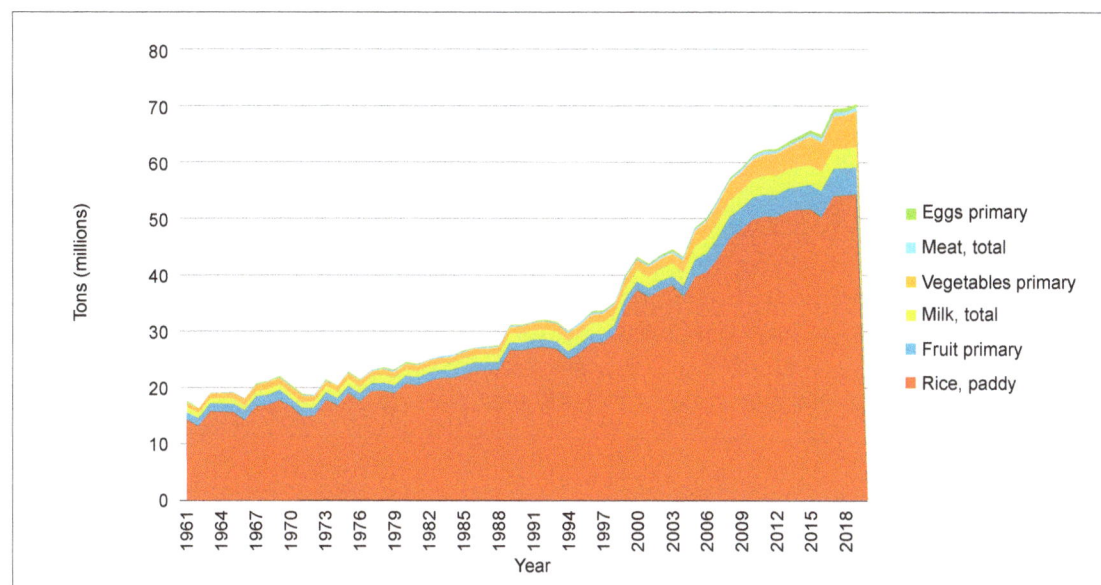

Source: Food and Agriculture Organization of the United Nations (FAO). 2021. FAOSTAT–Production. Crops and Livestock Products DatasetLPD (accessed 14 April 2021).

marginally poor are highly vulnerable to commodity price increases. Food security will be threatened if domestic food production is limited in the coming harvests. Increased prices for fertilizers, seeds, and energy may adversely affect agricultural growth, yielding unemployment and contributing to food insecurity. The country already has a significant trade deficit, and with the depreciating currency and other constraints, it may be difficult to import substantially more food if domestic supply is limited. Likewise, the safety net programs have limited reach, with the rural poor potentially highly vulnerable. Still, the government has responded to the food price crisis, for instance, by cutting customs tariffs and duty on rice imports. The government is also extending social programs such as offering commodity goods at subsidized prices to the poor through the Trading Corporation of Bangladesh.

In Bangladesh, irrigation is a vital input to produce crops, particularly rice, but yield gaps remain.[20] In the dry season, access to irrigation water plays a significant role in cultivating high-yielding varieties of rice and HVCs. The country's current cropping intensity[21] stands at around 194%. Despite improvement in recent decades, significant yield gap persists for most crops, including rice, largely due to limited adoption of suitable modern production technologies, inputs, and practices. Yields and food availability also vary significantly due to extreme weather events like seasonal floods, cyclones, and droughts.

---

[20]   The yield gap is the difference between actual farm yield and the potential yield with good management responding to agroecological conditions.

[21]   Cropping intensity is the number of times a crop is planted per year in a given agricultural area.

Bangladesh's irrigated area expanded rapidly after 1987 following the liberalization in the import of diesel engines, a reduction in import duties, and the withdrawal of the standardization of irrigation equipment. Until this time, Bangladesh Agricultural Development Corporation owned, operated, and maintained existing pumping sets.[22] These reforms led to private sector investment in minor irrigation equipment such as low-lift pumps, shallow tube wells (STWs), and deep tube wells (DTWs). The use of STWs and DTWs for extracting groundwater has expanded significantly. In 2019, about 1.5 million low-lift pumps, STWs, and DTWs were operated in Bangladesh, lifting groundwater to more than 19 million farmers for mainly dry season irrigation across approximately 5.3 million hectares.[23]

Crops are the main contributor of agriculture gross domestic product. Major crops are rice—Bangladesh is the world's fourth-largest producer—as well as jute, potato, wheat, oilseeds, vegetables, spices and condiments, pulses, and maize.[24] Production areas of pulses, oilseeds, fruits, vegetables, and spices are increasing, and total production has also increased in recent years (Figure 3). The central bank of Bangladesh provides subsidized loans to farmers at an annual interest rate of 4% to diversify production away from rice and expand the production of pulses, oilseeds, and spices along with maize and other crops.

Figure 3: Major Crops in Bangladesh, FY2009–2010 and FY2019–2020

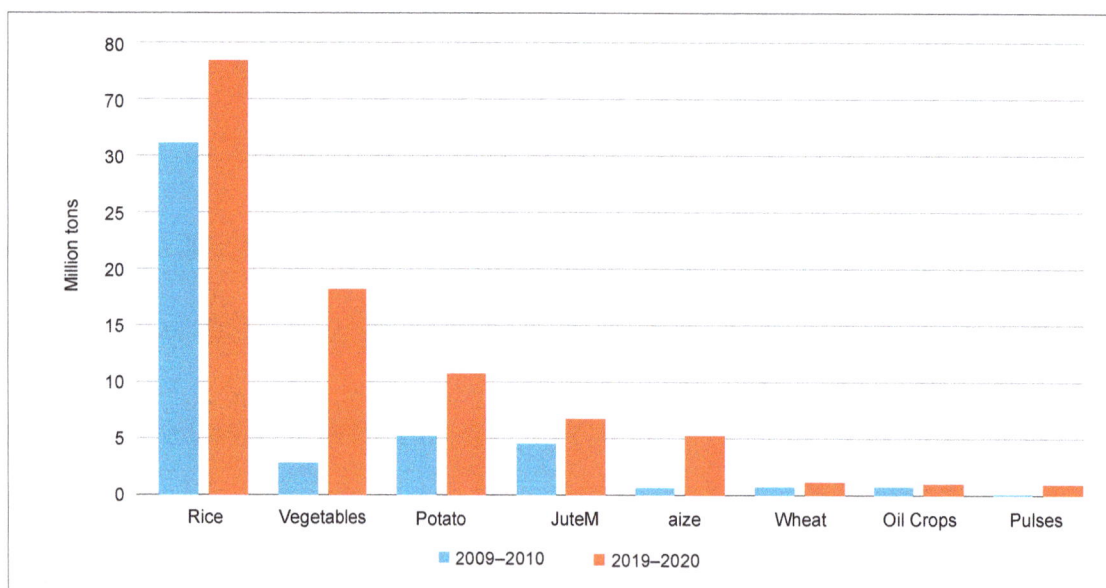

FY = fiscal year.

Note: Jute is in millions bell.

Source: Government of Bangladesh, Ministry of Agriculture.

---

[22] M. Hossain. 2010. *Shallow Tubewells, Boro Rice, and Their Impact on Food Security in Bangladesh.* In D.J. Spielman and R. Pandya-Lorch, eds. *Proven Successes in Agricultural Development: A Technical Compendium to Millions Fed.* Washington, DC: International Food Policy Research Institute.

[23] Bangladesh Agricultural Development Corporation. 2020. Summary of Irrigation 2018–2019. Dhaka.

[24] Food and Agriculture Organization of the United Nations (FAO). FAOSTAT: Countries by Commodity (accessed 22 October 2021). In 2019, the top three producers of rice paddy in the world were the People's Republic of China (209.6 million tons), India (177.6 million tons), and Indonesia (54.6 million tons). Bangladesh's rice production was just behind Indonesia also with 54.6 million tons.

Figure 4 illustrates the drought and flood hazards that affect rice, maize, wheat, and potato production. The hazards range from moderate to severe and are present in all districts. Some areas, in particular west Bangladesh, are prone to both risks. This figure shows that districts that are prone to severe flooding are dominated by rice production. HVCs tend to be produced in districts that either have a small proportion of areas affected by moderate flood, such as Gopalganj (southwest), or are more prone to droughts, such as Natore and Joypurhat (northwest). Therefore, the development of HVCs and their potential is closely interlinked with the need to adapt to these risks.

Products with high export potential. Floriculture, jute, coffee beans, and cashews are products with high export potential. Floriculture is a multibillion-dollar international business and expanding in Bangladesh.[25] Due both to increasing domestic demand and export potential, cultivation, marketing, and exports of different flowers are growing. More than 10,000 hectares are used for flower cultivation in the north and west of the country. Jute has been in increasing demand throughout the world as an environmentally friendly product. Aside from its potential for replacing plastics and polyethylene, jute is a stress-tolerant crop suited to Bangladesh's frequent periods of sustained drought. To meet a growing domestic and international demand for coffee beans and cashews, cultivation is increasing in Chittagong Hill Tracts and other areas. Although processing plants and facilities are yet to be established, the private sector is showing interest in these products. Challenges producers face in reaching their export potential are their small scale of production and comparatively lower quality. For all these products, there is a clear need to provide suitable value chains and support product development so that the country can take full advantage of growing demand.

Livestock development and production. Livestock raising has become increasingly commercialized in rural areas in recent years. However, household-level production remains predominant. About 20% of the population is directly involved in livestock production and about 50% of the population is indirectly involved through supporting inputs or marketing of livestock products. Many rural entrepreneurs, in particular women, have become involved in livestock-related enterprises due to its profit potential. Higher credit flow via nongovernment organizations (NGOs) and commercial banks in rural areas has contributed towards progress.

Furthermore, overall progress in the subsector has been encouraged by the introduction and expansion of artificial insemination and breeding programs across the country. However, the average productivity of a local cow—221 kilograms of milk per year—and average meat production of 50 kilograms per cow remain below the average of other developing countries.[26] Despite the significant increase of production of livestock products in recent years, domestic production of milk and milk products falls short of domestic demand, which is currently met by imports. Production data of key livestock products for 6 consecutive years is provided in Figure 5.

 The leather industry is closely linked with this subsector. Raw hides are supplied from cattle, goat, sheep, and buffalo raising. Leather industries have high potential for domestic as well as export markets, with finished footwear and other processed leather products already being exported. Black Bengal goat leather from Bangladesh is recognized as a high-quality leather.

During FY2019–2020, the contribution of livestock to national GDP was 1.60%, with an annual growth rate of 3.04%.[27] The livestock sector plays a key role in meeting dietary protein requirements for the population via supplying meat, milk, eggs, and other related products. In recent years, total meat production increased

---

[25]   Floriculture includes production of bedding and garden plants, potted flowering plants, foliage plants, cut flowers, cut cultivated green plants, and orchids.

[26]   World Bank. 2018. *Climate-Smart Agricultural Water Management Project (P161534)*. Washington, DC.

[27]   Government of Bangladesh, Ministry of Finance. 2020. *Bangladesh Economic Review 2020*. Dhaka. As per Bangladesh Economic Review 2020, the livestock subsector contributed about $5.27 billion or 12.83% of agriculture's contribution to GDP.

Figure 4: Estimate of Crops Affected by Droughts and Floods in Bangladesh

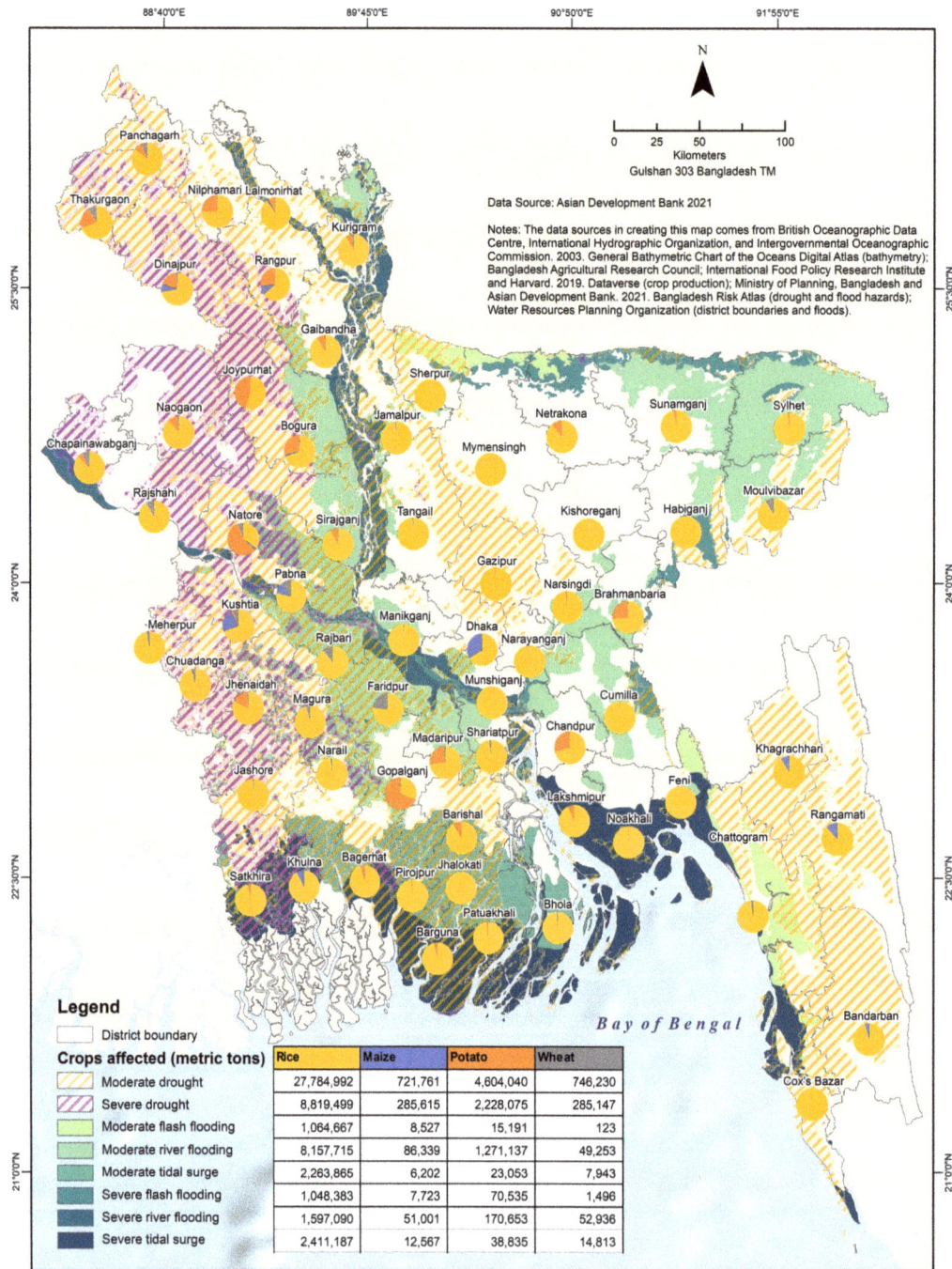

Crops affected (metric tons)	Rice	Maize	Potato	Wheat
Moderate drought	27,784,992	721,761	4,604,040	746,230
Severe drought	8,819,499	285,615	2,228,075	285,147
Moderate flash flooding	1,064,667	8,527	15,191	123
Moderate river flooding	8,157,715	86,339	1,271,137	49,253
Moderate tidal surge	2,263,865	6,202	23,053	7,943
Severe flash flooding	1,048,383	7,723	70,535	1,496
Severe river flooding	1,597,090	51,001	170,653	52,936
Severe tidal surge	2,411,187	12,567	38,835	14,813

Source: Asian Development Bank.

Figure 5: Livestock Production in Bangladesh, 2014–2020
(million metric tons)

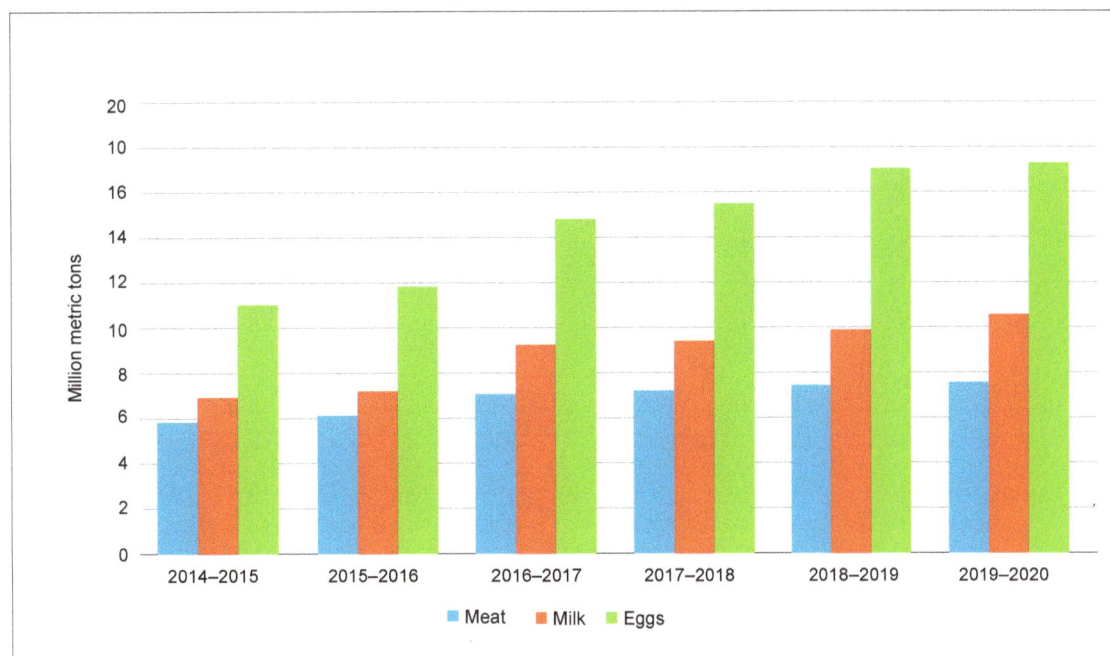

Source: Government of Bangladesh, Ministry of Fisheries and Livestock. *Bangladesh Economic Review 2020 and Annual Report 2019–20*. Dhaka.

significantly, from 5.86 million metric tons in FY2014–2015 to 7.51 million metric tons in FY2018–2019, an increase of 28.16%. This has made the country almost self-sufficient in meat production.

To meet the growing demand for meat, poultry, and eggs, there is a need to ensure the delivery of services such as artificial insemination, veterinarian health services, vaccinations, access to finance for farmers and entrepreneurs, establishing market linkages, and processing industries.

Fisheries development. Fish and shellfish provide about 60% of animal protein consumption in Bangladesh. Fisheries and aquaculture can be divided into three subsectors: marine fisheries, inland fisheries, and aquaculture. In 2018, Bangladesh had the third-largest inland water catch, and was the fifth-largest aquaculture producer in the world.[28] Fisheries are also an important source of exports. Demand for fish is growing and production in all three subsectors is increasing (Figure 6). In recent years, aquaculture has had the fastest growth, with significant potential to stay on this trend. In 2018, fisheries contributed 3.52% to national GDP and 26.37% to agriculture GDP. About 11% of the total population is engaged (directly or indirectly) in fisheries. The role of women is important, as they take on key responsibilities like drying, processing, and marketing, suggesting that they need special attention and tailored support to better integrate into fish value chain processes.

---

[28]   FAO. 2020. *The State of World Fisheries and Aquaculture 2020*. Rome.

Figure 6: Production of Fishery Products (million tons)

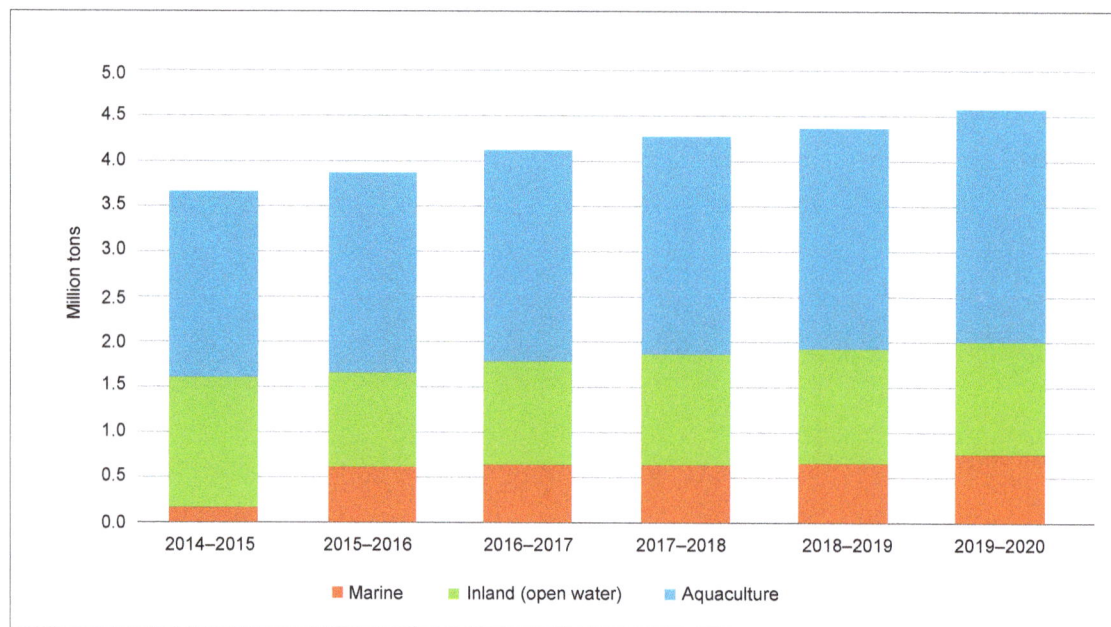

Note: Fact in fiscal year (FY) 2014-15 and FY2018-19; projection for FY2019–20.

Source: Government of Bangladesh, Ministry of Fisheries and Livestock. Bangladesh Economic Review 2020 and Annual Report 2019–20. Dhaka

Developing the "blue economy" through coastal and marine fisheries, including coastal aquaculture, represents an important opportunity for growth, contributing to food security and poverty reduction. Nineteen coastal districts represent nearly one-third of the country's land area and serve as gateways to ocean-based resources.

Fisheries and aquaculture face a range of challenges. Inland fisheries, while traditionally abundant, are low yielding and vulnerable to human activity and habitat degradation. The impact of increasing pollution loads on water quality is a concern. Increased irrigation and flood protection structures can impede migratory fish access to natural spawning grounds. Climate change is resulting in reduced fish catches due to higher temperatures, saltwater intrusion, and extreme weather events (drought and cyclones). Growth of aquaculture has increased demand for inputs, resulting in lower availability of quality seedstock (e.g., juveniles). There are limitations with the quality of other central inputs as well, such as quality feed and access to knowledgeable and timely extension services. Often without proper extension services, diseases can become a problem. Aquaculture also requires capital investments. To date, financing for this is generally only provided to larger producers. Marine fisheries face similar capital constraints to procure the equipment needed for deep-sea fishing. As such, much of the production is done by artisanal fishers in coastal and estuarine waters. For marine fisheries to grow, sustainable management plans are required along with systems that maximize the value of fish catches and ensure equitable benefits and sustainably developed coastal areas. An overarching problem across the sector are sanitary and phytosanitary trade barriers, which have limited export potential.[29]

---

[29]   Government of Bangladesh, Ministry of Fisheries and Livestock, Department of Fisheries. 2019. *Development Project Proposal. Sustainable Coastal and Marine Fisheries Project.* Dhaka.

C.   Agriculture, Natural Resources, and Rural Development Issues

The rapidly evolving context for agriculture and rural development in Bangladesh has significant implications for ANR. The growing and diversifying economy, with an expanding middle class, is increasing the demand for higher-value and more processed agricultural products. The garment industry and services sectors, which drive the economy, are diverting human and natural resources away from agriculture. At the same time, the natural resource base on which agriculture depends has been degraded and is vulnerable to climate change. This section explores the issues and challenges facing the sector in more detail; a problem tree presenting these and crosscutting sector issues is provided in Appendix 1.

Agriculture and Rural Development Transformation

The context for promoting greater agriculture and rural development is complex. A dynamic agriculture sector (including modernized value chains) and a growing nonfarm rural economy are key for sustaining structural transformation. Given the integrated nature of agriculture and rural development, advancing the sector requires significant capacity and coordination across sectors such as energy, finance, health, and education, and institutions with a significant role for private sector-led agriculture.

Global Food Security Index and malnutrition

The Global Food Security Index provides a useful overview of the opportunities and constraints for agricultural development (Figure 7). Food production, availability, affordability, quality, and safety depend on several factors, as presented earlier. The index incorporates indicators that measure food security across the world, including natural resources and resilience.[30] In 2020, Bangladesh ranked 21 out of 23 Asian countries, and 84 out of the 113 countries assessed. The country is considered to have high food security and access policy commitments, high marine biodiversity, with low prevalence of overexploited stocks, low volatility of agricultural production, and low food loss. However, it has an overall poor performance and is very far behind the world average for food affordability and dietary diversity, particularly in terms of micronutrient availability and protein quality. Overall food availability and affordability are impacted by low-quality roads and other transport infrastructure, and farmers' restricted access to finance. The index also considers vulnerability to droughts, floods, and storms and points to an extremely low presence of national agricultural adaptation policies and disaster risk management (DRM).

Despite making significant gains in tackling malnutrition, 28% of children under 5 years of age in Bangladesh are stunted.[31] Micronutrient deficiencies remain a major challenge. Diets are dominated by rice consumption with low intakes of fruits, vegetables, pulses, and meats. Intra-household food distribution is uneven, with sacrifices made by adolescent girls and women. About 70% of rural women and children do not meet minimum dietary diversity standards.[32] Similarly to many other ADB developing member countries, the cost of a healthy diet is too high for most of Bangladesh's population: 18.9 % of the population cannot afford a nutritious diet and 74.6% cannot afford a healthy diet, while almost all people in the country can afford an energy-rich diet.[33]

[30]   The Economist Intelligence Unit, Global Food Security Index (accessed 28 May 2021).

[31]   United States Agency for International Development. 2021. Bangladesh: Nutrition Profile.

[32]   Development Initiatives Poverty Research Ltd. 2020. Bangladesh Country Nutrition Profile.

[33]   FAO, International Fund for Agricultural Development, UNICEF, World Food Programme, and World Health Organization. 2021. *The State of Food Security and Nutrition in the World 2021. Transforming Food Systems for Food Security, Improved Nutrition and Affordable Healthy Diets for All.* Rome: FAO.

Figure 7: Summary of the Global Food Security Index for Bangladesh

Very Good	Moderate	Very Weak
• Volatility of agricultural production • Absortive capacity • Marine/fisheries biodiversity	• Access to finance (limited products) • Agriculture infrastructure (road weaker) • Sufficiency of supply • Food safety	• Nutrition, protein quality • Research and development expenditure • Land and water degradation • Adoption of climate resilience • Governance

Note: Bangladesh ranked 21 out of 23 Asian countries and 84 out of 113 overall.
Source: The Economist Intelligence Unit, Global Food Security Index. (accessed 28 May 2021).

Agriculture investment boosts productivity and distribution efficiency of food systems, improving nutrition security outcomes through enhancing rural income, as well as improving the cost structure of the food system, which together improves food affordability.

Demographic and economic trends and natural resource base

Several significant factors and trends provide a strong foundation for the continued growth and development of the agriculture sector. The large and still growing population provides a large domestic market for agricultural goods. In addition, rapid economic development and urbanization is increasing demand for higher value (e.g., fruits, vegetables, and animal protein) and processed and packaged products. The generally good soils and availability of water have allowed for a stable and growing supply of staple foods (e.g., rice). Marine and inland fisheries provide an accessible source of protein and can be further developed. Furthermore, the COVID-19 pandemic has drawn the attention of decision-makers to the importance of ensuring food and nutrition security, including through the availability of diversified, nutritious, and safe value-added food products to consumers. As post COVID-19 recovery measures are developed, measures to increase production, along with efforts to value add and process more foods, deserve high attention.

Research and development and extension services for agricultural development

The country's agricultural research and extension base has been particularly focused on paddy, which has brought the country to near self-sufficiency in rice. However, the share of agricultural researchers engaged in horticulture research is one of the lowest in the region. While high returns to agricultural research and development (R&D) are well documented, agricultural R&D investment in 2016 was 0.38% of agriculture GDP in Bangladesh, lower than Nepal at 0.42% and Sri Lanka at 0.62%. In addition, most R&D was conducted by public research institutions, with only 25% of agricultural R&D conducted in agriculture universities. Agricultural tertiary education institutions have human resources for R&D, with large spillover effects or externalities, but they do not receive adequate funding support and use most of the budget for salary of university staff rather than for

R&D. The limited R&D funding is also spent mostly in traditional agricultural sciences. Interdisciplinary research collaboration with nontraditional fields such as data science and robotics is rarely promoted, and collaboration between universities and industries is largely missing. Selective investment in agricultural tertiary education and research will help in harnessing technologies for higher productivity, supplying safe and high-quality food to domestic and international markets, and increasing the profitability and competitiveness of Bangladesh's agriculture in the global food market. While rice yields have improved, investment in R&D on increasing the productivity of other crops and livestock and aquaculture, where substantial yield gaps remain, is needed.[34] The government, through public research institutes and universities, plays a critical role in setting research priorities. Greater coordination and input from the private sector would also help focus public sector efforts in R&D.

Agricultural tertiary education

Along with R&D, tertiary agricultural education is crucial to enhance agricultural productivity. Unfortunately, Bangladesh's education sector, especially higher education, suffers from a lack of investment. Education expenditure was 2% of GDP in 2018, while university expenditure was only 1% of government expenditure. Bangladesh has six public agriculture universities with capacity to enroll approximately 15,000 students. The universities supply graduates to the agriculture sector largely focused on rice production. Increasingly it is recognized that graduates will need to be equipped with latest agricultural knowledge and skills, while facing the challenges of dwindling resources, climate change, and exogenous shocks like the COVID-19 pandemic. Future graduates will need to be able to accelerate the speed of technology adoption in biotechnology, remote sensing, weather analytics, data science, artificial intelligence, and robotics, and facilitate changes in the agricultural value chain. Agricultural universities face challenges in delivering quality education. Teaching and learning follow a teacher-centered approach with limited practical sessions and little integration of digital technologies for blended learning. During COVID-19 lockdowns, only 10% of the Bangladesh Agricultural University faculty continued teaching online, while the national tertiary education average was 56%. Assessment puts large weight on final written exams, and continuous assessment is seldom exercised. The curriculum was updated in 2019 after many years; however, it does not cover technology adoption in industry practices and remains constrained in what it can cover due to physical learning spaces and equipment.[35]

Agricultural productivity and access to finance

As already described, ANR in Bangladesh focuses on consolidating and expanding the productivity gains already achieved in food grain production. Nevertheless, potential remains to further increase yields. More recently, policy makers have directed more attention toward designing policies, strategies, and actions to accelerate crop diversification and commercialization by increasing local and export market opportunities for farmers and other stakeholders. The agriculture sector, largely focused on rice, has been a priority for public investment. The annual budget allocation and expenditure for the sector is growing, with $1.05 billion allocated in the FY2019–2020 budget for subsidies on fertilizer and other agricultural inputs, rising to $1.11 billion for FY2020–2021. Despite this increase, the availability of agricultural inputs for non-staple crops and HVCs remained stagnant. Government support, research, and extension are still heavily focused on rice. Crop diversification is promoted at the farm level, but access to quality seeds for HVCs remains a limitation, while much more support for improving the quality of extension services is needed. On-farm practices and systems need to become more efficient, product quality and safety needs improvement, and climate resilience and adaptation measures need to be better integrated.

---

[34]   World Bank. 2020. *Promoting Agri-Food Sector Transformation in Bangladesh: Policy and Investment Priorities.* Washington, DC.

[35]   ADB. 2021. *Strengthening Agricultural Tertiary Education Project in Bangladesh: Initial Poverty and Social Analysis.* Manila.

Limited access to finance is a significant constraint in developing and growing rural microenterprises and small agribusinesses, particularly for women in Bangladesh.[36] Microenterprises can promote rural economic growth by creating employment opportunities and linking to larger private sector actors, but they face nonfinancial and financial constraints. Microenterprises are financed largely by informal sources such as individual savings and informal loans from local moneylenders (who are often paid back by harvested crops at a predetermined low price along with cash interest in a practice known as *dadon*), friends, and relatives. Microenterprises and small entrepreneur successes have often been supported through microfinance and microfinance institutions. Lending to groups through Grameen Bank and other institutions, such as Bangladesh Rural Advancement Committee, a development NGO, have been successful in improving rural livelihoods, particularly when coupled with other social services (e.g., literacy). These interventions are also useful for balancing seasonal consumption and generating self-employment. The microfinance sector has also significantly focused on women's participation and empowerment. Unfortunately, few microenterprises receive funding from microfinance institutions or formal banks. In 2015, 85% of microenterprise investments were self-funded, 11% came from microfinance institutions, and just 4% came from banks. The overall microenterprise loan demand–supply gap was about $5.10 billion in 2015.[37]

Many of Bangladesh's agribusinesses are SMEs. Their investment needs are too large for traditional microfinance, but too small for banks. Many banks do not offer lending products that are sufficiently tailored in terms of amount or duration to meet their needs. Other financial products, such as crop insurance, are underdeveloped. Consolidated effort is needed from key stakeholders such as farmer groups, financial intermediaries, and regulators to identify and mainstream suitable products. Nonfinancial constraints include (i) lack of infrastructure facilities and utilities; (ii) inflexible regulations; (iii) a low level of corporate governance, i.e., less sophisticated counterparties; and (iv) local taxation laws that prevent more efficient organization structures. The potential and scope for on-farm-related SMEs to contribute toward agricultural development appears to be very promising. Designing and introducing new technologies, including digital financial services, will boost agricultural SMEs' decision-making and financial inclusion and improve their productivity and profitability.

Rural development and rural livelihood improvement

Increasing agricultural productivity and value addition is key to enhance the income of the rural population, thus reducing the proportion of the population living in poverty and inequality. Despite Bangladesh's substantial aggregate growth, small and often subsistence-oriented farms face many challenges to escape poverty. Although small farmers have often demonstrated creativity, resilience, and entrepreneurial spirit, particularly in the use of microcredit, the amount of income they can generate by growing primarily rice is limited. A further limiting factor is the small average farm size. The country's vulnerable hilly and low-lying areas, which have often been neglected, including coastal, *char*,[38] *haor*,[39] and wetland areas need long-term support and comprehensive interventions for inclusive development and livelihood improvement, taking into consideration the adverse impacts of climate change. These regions often also lag in terms of human development. In this context, human resources development, promoting rural organizations, and community capacity building should be priority areas for improving the social and economic conditions of the rural population. ADB is helping improve livelihoods in the ethnically unique Chittagong Hill Tracts through supporting institutional development and capacity

---

[36]   Women microentrepreneurs in agriculture face a different set of financial and nonfinancial constraints, including lack of business management, entrepreneurial and technical skills, access to start-up finance, information and networking support, and self-confidence.

[37]   ADB. 2021. *Microenterprise Financing and Credit Enhancement Project: Initial Poverty and Social Analysis.* Manila.

[38]   A *char* is a tract of land surrounded by the waters of an ocean, sea, lake, or stream. In the dynamics of erosion and accretion in the rivers of Bangladesh, the sandbars emerging as islands within the river channel (island *chars*) or as attached land to the riverbanks (attached *chars*), often create new opportunities to establish settlements and pursue agricultural activities on them.

[39]   A *haor* is a marshy wetland ecosystem in the northeastern part of Bangladesh which physically is a bowl or saucer-shaped depression that looks like an inland sea during monsoon floods.

building; providing rural access roads, markets, and village infrastructure; and assisting development of micro-agribusiness.[40]

Mechanization and technologies in agriculture and rural development

Farm mechanization plays an important role in the transition from traditional subsistence agriculture to commercial agriculture. Due to labor scarcity in rural areas and to increase productivity, mechanization practices are becoming more important and gaining popularity. Mechanization started with the use of DTWs and STWs for irrigation water of high-yielding varieties of rice in dry season (*Boro*). Currently, mechanized practices are mostly concentrated in rice and other HVCs on a limited scale. Areas of mechanization include land preparation, planting seedlings, harvesting, threshing, milling, and packing. Currently, machinery like tractors, pumps, and harvesters are rented as small-scale farmers do not have the resources to buy their own. Despite their popularity, the rental schemes have been slow to scale up throughout the country.

Mechanization is underdeveloped and hampered by a range of constraints. Small farms and fragmented lands do not provide economies of scale for farmers to individually invest in on-farm equipment. Limited access to finance inhibits farmers and small entrepreneurs in purchasing agricultural machinery. There is inadequate after-sales service for agri-machinery and services centers (including rental markets). Farmers and extension services have limited capacity in their abilities to use equipment. The local equipment market is underdeveloped, heavily reliant on expensive imports, and lacking skilled personnel to produce machinery and parts locally. Other disincentives to investing in machinery include the adverse effects of disasters and variations in load-bearing capacities of area-based soil.

Information and communication technology for agriculture development

The potential for ICT and digital technologies in Bangladesh to enhance the efficiency, climate resilience, and environmental sustainability of the agricultural production and food system is significant. For overall agriculture development, ICT can play a role in services delivery to producers or farmers and other actors in the production and food supply chain.[41] In addition, ICT may contribute toward automation by establishing area-wise databases for farmers, crops, production technologies, and weather data. ICT services became more important during the COVID-19 pandemic for maintaining food production, marketing, and providing food at the consumer level.

ICT and digitization is hampered in the ANR sector by (i) lack of coordination and collaboration among organizations under the ICT division, key ministries, and relevant organizations; (ii) lack of a unique farmers' database; (iii) farmers' limited access to smart ICT devices; (iv) lack of internet and mobile networks in remote areas; (v) the knowledge gap among farmers; (vi) a lack of awareness-raising campaigns at grassroots level; (vii) small farm size; (viii) inadequate services delivery by concerned organizations; and (ix) delayed data and information updates on pertinent websites.

Supply and value chains

With increasing demands for safe, nutritious, and fresh food as well as food processing, there is tremendous demand for integrated and efficient agricultural value chains. The rapid pace of urbanization and growth of supermarkets requires more integrated market and logistics infrastructure for greater value chain efficiency and improved sorting, packing, and storage facilities to ensure better quality. Also, the development of responsive rural institutions to help ensure that better market access brings improved livelihoods and social development is needed.

---

[40]   ADB. Bangladesh: Second Chittagong Hill Tracts Rural Development Project.

[41]   ICT provides access, for instance, to the latest knowledge about finance, high-yielding varieties of crops, fertilizers, irrigation water, integrated pest management, postharvest management, processing, and marketing.

Effective postharvest management, value addition, processing, and marketing are highly relevant and essential for increasing productivity and competitiveness of products. The main stakeholders are producers, entrepreneurs, value-adding and processing organizations, the government, and consumers. Ensuring win–win partnerships and efficient coordination among stakeholders is key for success. Yet to date, the interactions and collaborations between these actors are limited and lack vision towards the forward and backward linkages throughout the value chain.

The agricultural system in Bangladesh is still predominantly focused on the production of staple crops by smallholder farmers. Government incentives and subsidies have played a significant role in supporting and maintaining this system. Throughout the country, market demands are mostly dependent on supply systems. Farmers often sell at spot markets or directly to traders and brokers who are often syndicated and seek to maximize their profits at the farmers' expense. Thus, due to recurrent low prices for rice and other staple crops, farmers have little incentive to add value in terms of quality, sorting, or aggregation.

Inefficient and ineffective markets and agricultural value chains contribute to high postharvest losses in Bangladesh. These losses begin on-farm through poor and inefficient practices and are compounded by poor logistics and market infrastructure including a lack of cold chain facilities. Postharvest loss is a major problem across South Asia.[42] Around 42% of fruits and vegetables and 30% of grains are lost before reaching consumers.[43] By reducing fruit and vegetable postharvest losses, countries like Bangladesh could save up to almost $2 billion every year.[44]

There are not enough modern marketing, wholesale, and retail systems. Public marketing has been slow to evolve to meet changing demands. To fill the gap, the private sector has started engaging directly with farmer groups through contract farming and technical assistance to ensure the needed quantity and quality of production to support their processing.[45] In this way, the private sector is supporting value chain development and strengthening production. However, given the size of the agriculture sector in Bangladesh, these activities are limited as there are few large agribusinesses. Similarly, few agricultural public–private partnerships have been formed to date.

Food safety standards

Due to a general lack of quality assurance, food safety is neglected throughout the value chain (i.e., during production, processing, and marketing). Poor on-farm and postharvest management practices and a lack of food safety standards and testing procedures constrain the development of better quality assurance, while people engaged in agriculture and agribusiness have limited awareness of safe food practices. Food standards for more products must be developed, along with corresponding food safety regulations. More food safety inspectors and accredited testing laboratories are also needed.

Noncompliance or low-level compliance with Good Agriculture Practices Policy in the production systems of agricultural commodities, lack of hazard analysis, limited capacity to determine critical control points for agriproducts, and incomplete or noncompliance with other phytosanitary requirements for export are key impediments to increasing agriproduct exports.[46] Despite all these drawbacks, prospects and potential for diversified food processing and value addition are high in domestic and export markets.

---

[42] FAO. 2011. *Global Food Losses and Food Waste: Extent, Causes and Prevention.* Rome.

[43] FAO. 2020. *Fruit and Vegetables – Your Dietary Essentials. The International Year of Fruits and Vegetables, 2021, Background Paper.* Rome.

[44] ADB. 2020. *Dysfunctional Horticulture Value Chains and the Need for Modern Marketing Infrastructure: The Case of Bangladesh.* Manila; and ADB. 2019. *Dysfunctional Horticulture Value Chains and the Need for Modern Marketing Infrastructure: The Case of Viet Nam.* Manila.

[45] Examples of companies engaging in contract farming and technical assistance are ACI Limited, AGORA, MINABAZAR, INFINITY, Programme for Rural Advancement Nationally (PRAN), and SQUARE.

[46] Government of Bangladesh, Ministry of Agriculture. 2020. *Bangladesh Good Agricultural Practices Policy 2020.* Dhaka.

The lack of appropriate food safety standards and poor oversight of food safety regulations hinder the development of a strong private sector. Without enforced food safety standards, the private sector struggles to differentiate its products and enter export markets. Farmers also need to be familiar with the Good Agriculture Practices Policy to ensure that food safety measures are applied throughout the whole value chain. Furthermore, private sector companies struggle to find sufficiently skilled workers with the appropriate technical knowledge for their industries.

Water and Natural Resources

Integrated water resources management and environmental protection

Being home to the largest dynamic delta of the world and an extensive river network provides important benefits and challenges, mainly related to weather and climate change. During the monsoon, 55%–60% of the country is inundated during extreme flood events, while in the dry season, water has become even more acutely scarce than previously due to water diversion in upstream countries (footnote 12). Additionally, Bangladesh's coastal areas are prone to cyclones, storm surges, and sea level rise. Integrated water management gained a new relevance with the November 2018 approval of the Bangladesh Delta Plan 2100 (BDP2100), which establishes the best water management practices considering land use, agriculture, environmental management, and preservation of biodiversity to minimize the risks of natural hazards while protecting farm incomes and livelihoods.[47]

Water resources management is a high priority area for agricultural production and in supporting livelihoods of vulnerable rural communities. Flood control, drainage, supply of irrigation water, and integrated water management are the main components of overall water resources management. In addition, disaster risk management (DRM), conservation, and establishment of closer cooperation with riparian neighbors India, Nepal, Bhutan, Myanmar, and the People's Republic of China are strategic priorities for the sector. Over the last 3 decades, large flood control, drainage, and irrigation schemes, utilizing mostly surface water for irrigation, and costing $3.0 billion, have contributed to significant increases in agricultural production and increased farm and off-farm income (footnote 45).

Despite these past investments being focused on infrastructure development for water resources management, there has been limited success in improving institutional performance and deficient operation and maintenance (O&M).[48] One of the consequences of this is that large irrigation systems in Bangladesh suffer from inefficient water utilization and poor maintenance, and many needed repairs are backlogged. The low performance will worsen in the future due to the exacerbation of extreme weather events caused by climate change. Such risks need to be considered when rehabilitating infrastructure and modernizing system operations. In addition, more effort is required to improve community participation, particularly from women, and the involvement of the private sector in the sustainable management of water resources.

Flood management and disaster risk management

Bangladesh is one of the most flood-prone countries in the world, with different hot spots and hydrological regions facing different types of floods, including river flooding, flash flooding, rainfed flooding, waterlogging, and coastal flooding at different times of the year (footnote 12). The threat of frequent floods and riverbank erosion disasters discourages investment and leads to lower economic growth of riverine areas. Poverty is extensive in the areas

---

[47] Government of Bangladesh, Planning Commission, Ministry of Planning. 2018. *Bangladesh Delta Plan 2100: Baseline Studies Volume 1. Water Resources Management*. Dhaka.

[48] ADB. 2019. *Report on Performance Evaluation of ADB Agriculture, Natural Resources, and Rural Development Assistance Program 2009 to 2018 to Bangladesh*. Manila.

exposed to floods and riverbank erosion disasters, with districts more exposed to floods showing poverty rates above the national average (footnote 45). Climate change is also exacerbating these weather events by increasing rainfall during the monsoon and reducing rainfall during the dry season. Cyclones and storm surge events threaten especially flat and unprotected coastal areas in the south. With the acceleration of sea level rise due to global warming, a large area of agricultural cropping land will be submerged in coming years. Flood modelling predicts that the total flood-affected area will increase in the coming decades.[49] Additionally, the rising sea level will increase intrusion of saltwater into groundwater resources. Related to reducing river floods and managing erosion is the need for increasing waterway navigability. These are interconnected objectives, particularly given the braided nature of the river systems, and should be managed jointly.

Communities and assets can be further protected from flood events when they have access to flood early warning systems. The existing system needs to be updated to provide increased lead time, but this cannot be achieved without cooperation between institutions responsible for hydrology, meteorology, modelling (institutes and centers), dissemination, and early actions. Additionally, DRM must include investments to reduce the underlying risks and increased institutional capacity to promote new farming practices for greater adaptation to the effects of climate change.

Water for irrigation and modernization

Bangladesh's total irrigated area almost doubled between FY2000 and FY2016, when it covered 90% of total cultivated land. This was crucial to increase rice production, mainly Boro. However, this growth trend has since stagnated. Most of this area is fed by STWs and DTWs, which despite high recharge rates of aquifers, contribute to the decline of the country's groundwater table, especially in the northern districts (Figure 8). Groundwater accounts for 79% of irrigation, while the rest comes from surface water pumped from canals and rivers using low-lift pumps. Unfortunately, the conveyance of the extracted water is not done efficiently, with water losses of 60%–70%. The growth of irrigated areas has been associated with liberalization of the market for import of diesel motors, mentioned above. To reduce dependency on fossil fuels for irrigation and their associated carbon dioxide emissions, the installations of solar panels for small-scale irrigation in rural areas was identified as a priority in the most recent national strategies. It has also been highlighted as an area for potential private commercial investment. Notwithstanding this, developing strong water accounting models and a shift to volumetric water pricing might further incentivize water-saving techniques and crops.

Increasing storage capacity in existing surface water retention bodies and groundwater recharge technologies are needed to respond to the increasingly extreme weather events. Additionally, the modernization of systems is necessary to increase the efficiency of the conveyance of water for irrigation. There has been limited use of materials such as plastic, PVC or polythene pipes, buried pipes, and technologies for increasing on-farm water use efficiency, such as drip irrigation and fertigation for non-cereal crops. Precision agriculture also represents an opportunity to increase efficiency not only of irrigation water, but also other inputs for agriculture.[50] Smart irrigation and weather advisory services based on satellite data are starting to be used in the region. Using widespread services, like text messages, these advisory systems are reported to not only improve irrigation efficiency but can also be combined with flash flood warnings and allow for better risk-based decision-making about early harvests.

---

[49]   The inundation area due to climate change will increase by 6% in the decade following 2030, and 14% in the decade following 2050, compared to a base year of 2005.

[50]   Precision agriculture is the science of improving crop yields and assisting management decisions using high technology sensor and analysis tools. It employs data from multiple sources to improve crop yields and increase the cost-effectiveness of crop management strategies.

Figure 8: Groundwater Table Depth in Northern Bangladesh, 1982–2016

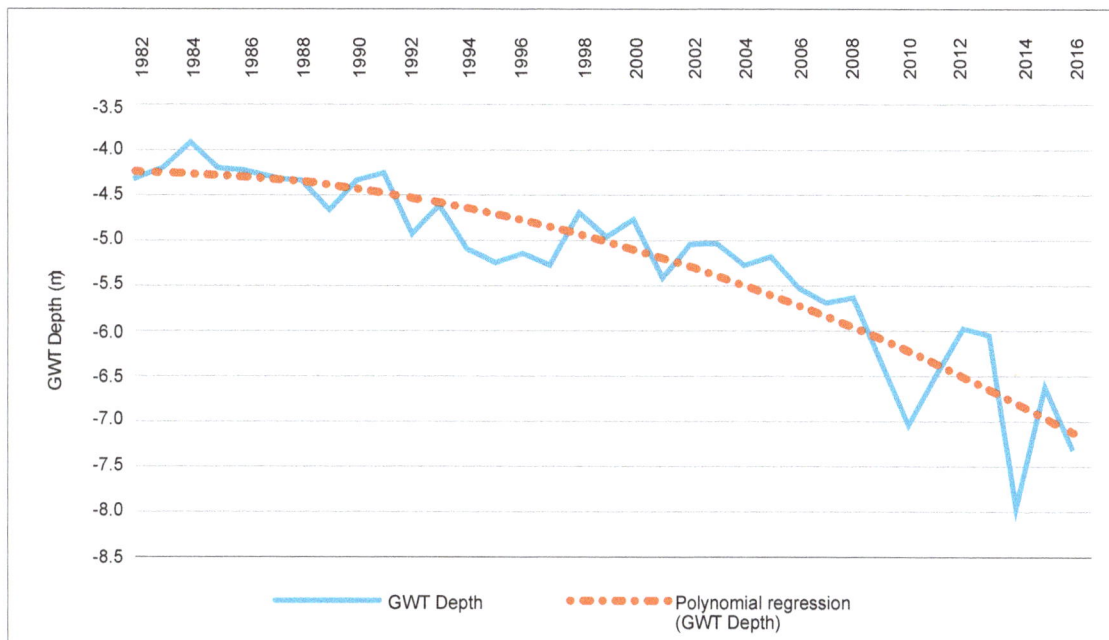

GWT = groundwater table, m = meter.

Source: Government of Bangladesh, Planning Commission. 2020. *Eighth Five-Year Plan, FY2021–FY2025.* Dhaka.

Saltwater intrusion and erosion in coastal regions

Due to sea level rise, the salinity frontier of the country will move upstream gradually in the coming years. Currently, saline groundwater can be found up to 100 kilometers inland (footnote 25). Excessive salinity currently affects 40% of Bangladesh's coastal area during the dry season. It reduces the soil's productivity and poses a risk to drinking water sources. To decrease the affected area to 30%, it will be important to ensure that the flows of rivers during the dry season do not fall below designated levels and to protect the coastline from erosion.

Being located downstream of the Ganges–Brahmaputra–Meghna basins, the flows that reach Bangladesh during the dry season depend on upstream countries' water abstraction. Presently, only 15% of the total transboundary flow is available during the dry season, which aggravates water shortages. The reduction of freshwater flows in the rivers also allows for the infiltration of brackish water upstream, pushing saltwater intrusion further inland. Therefore, enhanced regional cooperation for the management of shared water resources is crucial. Additionally, arsenic contamination of shallow groundwater hinders the utilization of groundwater in Bangladesh.

Polderization along the coastline started in the 1960s to control floods in low-lying areas and slow down saltwater intrusion.[51] However, these structures require either repair or redesign to remain effective in the face of climate change. Due to lack of proper maintenance and changing weather conditions, the polders have become dilapidated and are reducing the natural drainage of surrounding areas, thus increasing the risk of flooding and waterlogging in some of the polders themselves. The complex hydrodynamic environment that defines Bangladesh's rivers involves land erosion and accretion in certain areas, especially the Meghna estuary (footnote 45). Due to population pressure and lack of alternatives, vulnerable populations are moving into recently accreted areas, or *chars*, and robust planning for the occupation of these areas is required. Early occupation of *chars* hampers the vegetation growth that would support the consolidation of its soil and may put the population at risk.

Wetland, forestry, and habitat degradation

Bangladesh has a variety of wetland ecosystems, including the world's largest mangrove forest Sundarbans, *haors*, and oxbow lakes, which are semi-closed water bodies often found in floodplains. These ecosystems, estimated to cover about half the country's land area, play a crucial role in maintaining ecological balance and flood management. However, human intervention is endangering these ecosystems, which provide a critical source of income and nutrition for millions of people living in Bangladesh's rural areas. Despite efforts to protect its wetlands, progress is lagging, and these sensitive areas are still under threat (footnote 12).

Forests account for 14.1% of Bangladesh's land area.[52] In recent years, forestry policy has shifted from the exploitation of resources to conservation and sustainable management of ecosystems. The national priority targets associated with United Nations Sustainable Development Goal (SDG) 15 are to enhance forest area as a proportion of total land area to 18% and increase the area of tree-covered land by 25% in relation to total land area, starting from the 2016 baseline of 22%.[53] Forests help mitigate against storm surges, stabilize the environment, reduce the impact of earthquakes, develop healthy microclimates, manage watersheds, encourage ecotourism, and capture carbon. Demand for wood is growing at a faster pace than supply can meet, which will encourage illegal felling and imports. The Bangladesh Forest Inventory carried out a survey that found only 1.3% of respondents received support for sustainable tree and forest management from any organization, which might compromise sustainable forestry practices and incentivize illegal felling (footnote 52). According to the Eighth Five-Year Plan (FY2021–2025), forestry contributes 1.6% of GDP and 12% of agricultural GDP, but the value of non-timber forest products such as bamboo, rattan, nipa palm, and medicinal plants are starting to gain more recognition (footnote 12). Non-timber forest products hold important traditional value in the cultures of ethnic communities. Four tree species found in the Hill zone have been identified as threatened on both the International Union for Conservation of Nature and Bangladesh lists, while one mangrove species that occurs in the Sundarbans is critically endangered.[54]

Healthy soils are the backbone of biodiversity and crucial for carbon sequestration. The 2019 Bangladesh Forest Inventory estimates that the land owned by the Forestry Department holds 24.5% of the country's total aboveground carbon stock, with the main carbon stocks in Chittagong and Khulna. The rest of the carbon stock

---

[51]   Polderization is the reclamation of land from the sea or in wet zones by building levees, filling, and draining. Polders will be below the water level during part of the time, which means they are often at risk of flooding, and the surrounding dikes or barriers must be well maintained.

[52]   Government of Bangladesh, Forest Department, Ministry of Environment, Forest and Climate Change. 2019. *Tree and Forest Resources of Bangladesh: Report on the Bangladesh Forest Inventory*. Dhaka. Hill forest accounts for 4.6% of the country area, followed by shrubs with scattered trees (4.2%), and mangrove forest (2.7%).

[53]   SDG 15 is to protect, restore, and promote sustainable use of terrestrial ecosystems, sustainably manage forests, combat desertification, halt and reverse land degradation, and halt biodiversity loss.

[54]   The five zones of the Bangladesh Forest Inventory are Coastal, Hill, Sal, Sundarbans, and Villages. The Hill zone represents hilly geographic areas in the eastern part of the country.

is located within multiple and single crops, and rural settlements. However, with 65% of its land considered degraded, Bangladesh has the second-highest land degradation percentage among Asia and Pacific countries.[55] Interventions like nature-based solutions (NBS) are likely to provide more sustainable alternatives for water resources management and DRM than traditional "hard" infrastructure solutions, and provide additional services to the communities, such as protection and restoration of ecosystems, food, timber, increased biodiversity, and tourism.[56]

The concept of a circular economy associated with NBS has the potential to reduce the need for chemical fertilizers, reduce water pollution due to agrochemical runoff, and regenerate soils by promoting the return of nutrients that have been used for food production to farm soils. This can be done by processing fecal sludge and municipal solid waste for use as compost for soil conditioning and improving crop yields.[57] In addition, the reuse of these waste streams can provide opportunities for recovery of the O&M costs of treatment plants and other waste management infrastructure.

Crosscutting Sector Issues

Policy, regulatory, and institutional environment

The ANR sector needs to address structural and emerging challenges to ensure that agricultural production can equitably meet market demands. Policies and incentives, traditionally focused on rice self-sufficiency, need to evolve as demand for high-value commodities, such as vegetables, fruits, dairy products, meat, and eggs, increases. Subsidies for fertilizer and credit have supported rice production, which remains a priority. These inputs and others such as seeds and access to mechanization will be critical areas to address, particularly given limitations of land availability. To address these production challenges and increase value addition would require substantial transition in the roles of government institutions and corresponding policies and regulations to increase productivity and encourage more investment. In recent years, the government has introduced a range of new policies, but these require resources and at times accompanying regulations to modernize the sector and empower the private sector.

Bangladesh will also require measures to address the challenges of climate change, given the high vulnerability of smallholders to floods, droughts, and other extreme weather events. Improving resilience and adaptation, along with mainstreaming of disaster early warning systems, will be required. In addition, policies require institutional changes, particularly greater decentralization to empower and strengthen the local levels of government to better respond to these challenges. The current institutional arrangements are often siloed and do not facilitate the needed coordination and integration to develop a modern food system in Bangladesh. Furthermore, policies should reflect and support the nature of Bangladesh agriculture by being inclusive and responsive to small farmer and gender needs.

Rural electrification

Rural electrification is a key factor for agriculture and rural development. About 50% of installed power generation capacity has private sector investment. Transmission and distribution companies serving the capital city have transformed into companies listed in the market to raise capital for investment needs. The rural

---

[55]   UN Economic and Social Commission for Asia and the Pacific. 2021. SDG Gateway Asia Pacific. (accessed 6 May 2021).

[56]   NBS are inspired and supported by nature. They are cost-effective, simultaneously providing environmental, social, and economic benefits and helping to build resilience. Such solutions bring more, and more diverse, nature and natural features and processes into cities, landscapes, and seascapes, through locally adapted, resource-efficient, and systemic interventions.

[57]   Fecal sludge refers to the material accumulated at the bottom of septic tanks, pit latrines, or vaults, and consists mainly of fecal solids and urine. It differs from conventional wastewater due to its low water content and higher oxygen demand and suspended solids concentration.

electrification subsector, which serves low-income customers, adopts the consumers' cooperative model. This sector has been striving to decarbonize and is committed to help meet Bangladesh's Nationally Determined Contributions by improving energy efficiency and developing domestic renewable energy resources.

Since 2009, installed power generation in Bangladesh has tripled. However, as of 2016, 31% of the population in rural areas still did not have access to electricity, compared to 24% nationally. Just under 80% of existing power-generation capacity is operational, with frequent scheduled blackouts. On average, electrical outages can occur 80 times per month, lasting from 5 hours for businesses to 10 hours for households, which is particularly challenging for manufacturing and food processing. Bangladesh ranked 108th in terms of electricity access among 141 countries and 68th in terms of electricity supply quality in the 2019 Competitiveness Index of the World Economic Forum.[58]

The electricity price for agriculture is subsidized, primarily at the expense of industrial and individual users. The current electricity rate for farmers is 25% lower than the cost of electricity supply itself and it is lower than for individual users and businesses ($0.03 per kilowatt-hour for farmers against $0.04 for individual consumers and $0.08 for businesses). This has encouraged the expansion of DTWs and STWs for irrigation. While the availability of water has contributed to advances in food production, in some places it is degrading and contributing to the decline of groundwater tables due to over-withdrawals.[59] Gas-based power is also subsidized for fertilizer producers. Use of solar power in agriculture is limited.

Market connectivity—rural roads and transportation

In production and marketing systems of agricultural commodities, market connectivity has a multitude of impacts on lives and livelihoods especially for those who reside in remote and vulnerable areas. Connectivity and transportation are the backbone of all agricultural supply chains. Along with postharvest processing mentioned previously, they are essential to minimize postharvest loss, especially perishables and items with short shelf lives. Roads and inland water transport ensure 90% of the traffic generated in the country (footnote 12) and are essential for the commercialization of agricultural products. Besides increasing agricultural incomes, rural roads contribute significantly to employment opportunities while providing access to economic and social services to the entire rural population.

Rural connectivity and market access remain as challenges. Just over 60% of Bangladesh's population lives in rural areas, of which only 67% were served by all-season roads in 2016, according to the Rural Access Index.[60] Poor market connectivity, especially in rural areas, increases transportation costs and forces farmers to sell their products to brokers, leading to low profit margins for themselves. According to official records, there are 13,949 markets in Bangladesh, ranging from small village primary markets to large wholesale urban or metropolitan markets.[61]

[58]   World Economic Forum. 2019. *The Global Competitiveness Report 2019.* Geneva.

[59]   A. Zahid and S. R. U. Ahmed. 2006. *Groundwater Resources Development in Bangladesh: Contribution to Irrigation for Food Security and Constraints to Sustainability.* International Water Management Institute.

[60]   A. Iimi et al. 2016. New Rural Access Index: Main Determinants and Correlation to Poverty. *Policy Research Working Paper.* No. 7876. World Bank; and Azavea, Research for Community Access Partnership, and UK Aid Direct. *Rural Access Index* (accessed 4 May 2021).

[61]   Government of Bangladesh, Department of Agriculture Marketing. *Market Directory* (accessed 4 May 2021).

Urban–rural connection

Cities are resource intense and are increasingly becoming sinks of resources from rural environments such as water, food, and nutrients that are transformed into waste that needs to be disposed of. Interest in developing models of circular economies that will keep the value of resources throughout the economic cycle is increasing. Although agriculture and waste management might not appear obviously related, circular economy models are crucial to avoid the depletion of rural resources. Therefore, besides operationalizing its own waste management plans, the government is also incentivizing the industry to pursue the circular economy model by reusing and recycling materials, which can provide opportunities for rural areas by expanding business operations and closing nutrient and water loops.

Gender issues and women empowerment

Bangladesh has made significant progress in reducing poverty, improving equal access to quality health and education, and improving access to basic services for all. Despite these achievements, there is still inequality and various forms of exclusion associated with gender, age (older people, youth, and children), disability, social identities, sexual and gender identities, geographic location, and income status (footnote 1).

Bangladesh ranks 71st (out of 146 countries) in the Global Gender Gap Index, which benchmarks the current state and evolution of gender parity across four key dimensions (economic participation and opportunity, educational attainment, health and survival, and political empowerment), and is the best performing country in South Asia, having closed 68% of its gender gap.[62] Nevertheless, women's participation in the economy is one of the lowest in the world. Further, violence toward women in Bangladesh is still prevalent, with 53.3% of women having experienced violence in their lifetime, and 45.2% of women aged 15–19 married as children.

In addition, women are disadvantaged on account of:

(i)   low asset ownership, with only 13% owning agricultural land alone or jointly, compared to 70% of men in rural areas;[63]

(ii)   low labor market participation, particularly in the formal economy, including a large difference between participation rates of women (36.3%) and men (80.5%) (footnote 10);

(iii)   a high proportion of rural women being engaged in agriculture (68.1% in 2005–2006) compared to men (48%);[64]

(iv)   households headed by women accounting for 12.5% of all households in 2014, and these households being poorer, more socially excluded, and therefore more vulnerable;[65]

(v)   representing only 38.5% of the workforce overall but approximately 50% of the labor in agriculture, making them, and other vulnerable populations, disproportionately affected by climate change;

(vi)   being more likely to be landless; and

(vii)   their income being more likely to come exclusively from agriculture, so their livelihoods being more prone to climate change impacts.

[62]   World Economic Forum. 2022. *Global Gender Gap Report 2022* – Insight Report. Geneva.

[63]   A. Kotikula and J. Solotaroff. 2019. *To Gain Economic Empowerment, Bangladeshi Women Need Equal Property Rights*. World Bank Blogs. 1 May.

[64]   H. U. Amed. 2018. *Women's Contribution to Agriculture*. The Financial Express. 24 August.

[65]   World Bank. World Bank Data (accessed 24 August 2021).

This makes the participation of rural women in community decision-making committees such as water management community associations crucial for them to work towards equal benefits of interventions for their development.

Coronavirus Disease Impacts

In the short term, the COVID-19 pandemic seriously hampered farmers' market access for selling products, procuring raw materials, and hiring labor. In particular, productive capacities of poultry and dairy fell. Sales of flowers in the domestic market were reduced significantly as social gatherings were reduced. Emerging and relatively immature value chains, such as those for cut flowers, may need extra assistance to regain their momentum. A potential longer-term and more significant impact is increasing commodity and food prices, which are likely to remain high in the coming years. This will adversely affect the poorest and most vulnerable, increasing the food insecurity rates. These vulnerable people are also likely to be among the last to be vaccinated in Bangladesh. ICT has been an area of significant promise and growth during the pandemic. Its use has accelerated to connect farmers with consumers through e-commerce and to allow mobile financial services for trade.

II. Country Sector Strategy

A. Government Sector Strategies

In the last few years, the Government of Bangladesh has developed several strategic plans, policies, and laws that identify ambitious goals for future decades. All strategies recognize the importance of modernizing agriculture, developing further resilience to climate hazards, and sustainable management of natural resources. In addition, these policies also translate the government's commitment to attain the SDGs.

Perspective Plan 2041

The Perspective Plan of Bangladesh 2021–2041 aims to eradicate poverty from the country and attain high-income country status by 2041.[66] The plan recognizes that conscious management of key natural resources like land, water, forestry, natural habitats, and air is crucial to avoid their depletion and degradation. The plan also acknowledges that the modernization of agriculture, fisheries, and livestock, through diversification and climate resilience is essential to achieve food security and nutritional balance. This structural transformation is dependent on the development of human capital and skills that match current and future needs of the country.

Bangladesh Delta Plan 2100

The BDP2100 is a long-term holistic strategy to guide the country's water management until the end of the century.[67] The plan identifies the importance of water management for urban and rural areas and all economic sectors, with a strong focus on agriculture, and analysis of several climate change impact scenarios. This long-term plan sets out six national goals:

(i)   ensuring safety from floods and other climate change-related disasters;

(ii)   enhancing water security and efficiency of water usage;

(iii)   ensuring sustainable and integrated river systems and estuaries management;

(iv)   conserving and preserving wetlands and ecosystems, and promoting their wise use;

(v)   developing effective institutions and equitable governance for in-country and transboundary water resources management; and

(vi)   achieving optimal and integrated use of land and water resources.

---

[66]   Government of Bangladesh, Ministry of Planning. 2020. *Perspective Plan of Bangladesh 2021–2041*. Dhaka.

[67]   Government of Bangladesh, Ministry of Planning. 2018. *Bangladesh Delta Plan 2100: Bangladesh in the 21st Century (Abridged Version)*. Dhaka.

Importantly, the plan also includes an investment plan that consists of 65 physical projects and 15 institutional and knowledge development projects. Its total capital investment cost is $34.78 billion. All the projects are envisaged to start by 2026 and will extend over several decades, with opportunities for external investment also clearly defined.

Eighth Five-Year Plan

Every 5 years, the Government of Bangladesh prepares a plan that identifies the most pressing issues for the short-term development of the country. In December 2020, the Eighth Five-Year Plan 2020–2025 (footnote 10) was published.[68] In conformity with the Perspective Plan of Bangladesh 2021–2041 and BDP2100, it focuses on scaling up existing good practices of water conservation and management, integrated water management (including flood control and prevention schemes), early warning systems for flood and cyclones, and irrigation improvement (including efficient management of natural resources and livelihood improvement in coastal areas). For agriculture, the strategic priorities take a holistic approach, while ensuring food security and nutrition through increasing productivity,[69] minimizing yield gaps, stabilizing prices of agricultural commodities, taking measures to improve farmers' profitability and management of financial risks, diversifying HVC production, value addition, strengthening agricultural supply chains, increasing fisheries[70] and livestock production, increasing commercialization through digital technologies, easing credit facilities for small farmers, escalating human resource development for undertaking frontier research and delivery services, and ensuring efficient utilization of natural resources. The plan launched a series of innovative programs for rural development, such as One Village, One Product, which aim to create local employment, strengthen rural ecology, and rejuvenate local economies.

National Agriculture Policy 2018

The main goal of the National Agriculture Policy is to achieve safe and profitable agriculture, sustainable food production, and food and nutrition security.[71] It has the objective of improving socioeconomic conditions by increasing productivity and production of crops, diversifying crops, improving marketing systems, and ensuring profitable agriculture and efficient utilization of natural resources. To implement the policy, the government adopted in 2020 a plan of action, which focuses on improving the livelihoods of small farmers and intends to promote sustainable production and commercialization of agriculture to further reduce food insecurity and improve nutrition. However, to meet these objectives, significant investment and changes in practices, policies, institutions, and incentives are required. The action plan provides an important framework to guide government decisions and development partner support to the sector.

Water Policy

The National Water Policy, adopted in 1999, sets out a comprehensive framework for the water sector and for large surface water irrigation schemes. The government has established policy, legal, institutional, and planning frameworks for the water sector, which provide a suitable environment for developing necessary sector reforms. The 2013 Water Act revised and consolidated existing laws that govern the ownership, utilization, and financial management of water.

---

[68]   The government's Sixth and Seventh Five-Year Plans emphasized development and extension of appropriate production technologies and mechanized cultivation; diversification in both crops and non-crops; sustainable agricultural growth through more efficient and balanced utilization of land, water, and other resources; adaptation to climate change; establishing connectivity to markets; and river management, including flood control.

[69]   The government recently started a follow-up project (2020–2025) providing $35.27 million for the promotion of agricultural mechanization. Under the project, the government provides a 50% subsidy to farmers throughout the country and a 70% subsidy for those in *haor* and low-lying *char* areas.

[70]   The plan has a target of increasing fisheries production to 9% by 2025, with measures being taken to increase participation of women to at least 30% in aquaculture production and processing industries.

[71]   Government of Bangladesh, Ministry of Agriculture. 2018. *National Agriculture Policy - 2018*. Dhaka.

Other Policies and Laws

To further operationalize these strategic plans, several policies and laws have also been approved. Some of the most relevant to ANR include National Agricultural Extension Policy 2020, Bangladesh Good Agriculture Practices Policy 2020, National Organic Agriculture Policy 2016, Ground Water Utilization in Agriculture Policy 2019, Marine Fisheries Ordinance 2020, National Women Development Policy 2011, and Animal Welfare Law 2019. These laws and policies testify to the government's efforts to transform the ANR sector by improving its performance while highlighting the importance of the environment for the country. The identification of the main public stakeholders affecting the sector is presented in Box 1.

As ANR is highly vulnerable to the impacts of climate change, the government, supported by the World Bank, developed its Climate Smart Agriculture Investment Plan.[72] The plan identifies investment opportunities that are aligned with Bangladesh's Nationally Determined Contribution and BDP2100. It outlines five integrated investment packages relating to productivity, resilience, and mitigation.[73]

Although not directly related to agriculture, the Bangladesh Standards and Guidelines for Sludge Management 2015 and the Institutional and Regulatory Framework for Fecal Sludge Management 2017 enable the development of circular economy models to deal with organic waste, as described above. The Eighth Five-Year Plan highlights the Sakhipur fecal sludge treatment plant and co-composting plant, heavily supported by the Department of Agriculture Extension, as a model to be replicated across the country.

Box 1: Identification of Main Public Stakeholders

Bangladesh's agriculture, natural resources, and rural development sector is under the mandate of the Ministry of Agriculture; Ministry of Water Resources the Ministry of Local Government, Rural Development and Co-operatives; Ministry of Fisheries and Livestock; and Ministry of Environment, Forestry, and Climate Change.

Other relevant ministries are the Ministry of Food, Ministry of Commerce, Ministry of Industries, Ministry of Education, Ministry of Finance, Ministry of Youth and Sports, Ministry of Land, and Ministry of Disaster Management and Relief. The main responsibilities for the development and management of water resources are entrusted to the Bangladesh Water Development Board under the Ministry of Water Resources, whereas small-scale irrigation and flood control schemes are the responsibility of the Local Government Engineering Department, under the ministries of local government and agriculture. Other public organizations involved in water management are the Water Resource Planning Organization, National River Conservation Commission, and Bangladesh Inland Water Transport Authority.

---

[72] World Bank. 2019. *Bangladesh Climate Smart Agriculture Investment Plan: Investment Opportunities in the Agriculture Sector's Transition to a Climate Resilient Growth Path.* Washington, DC.

[73] The five packages are (i) Agricultural Innovation System, (ii) Gender-Sensitive Development of Homestead Production, (iii) Resilience Through Diversification, (iv) Livestock Upstream Value Chain Development, and (v) Climate-Resilient Agri-Livelihood Development (*hoar* areas).

B.   ADB's Agriculture, Natural Resources, and Rural Development Sector Support Program and Experience

ADB's assistance program for the sector has expanded and evolved over the years. Until the 1980s, the emphasis was on increasing food production through expanding farmers' access to inputs such as fertilizers and improved seeds. The focus broadened during the 1990s to include crop production, livestock, rural credit, social forestry, and related water resources management and rural infrastructure development. The interventions fostered access to economic opportunities including jobs, crop diversification, agribusiness, natural resources management and conservation, and DRM.

Over the last decade, ADB's assistance has focused more on infrastructure and supporting the enabling environment for agricultural growth, which are areas where ADB has a positive performance record and where other external assistance has been limited. Investments have concentrated on rural roads and water-related infrastructure for irrigation, flood protection, and DRM. The two most recent country partnership strategies (CPSs) reflected the changing corporate priorities within ADB. To increase private sector engagement and operations in the ANR sector, supporting market linkages to strengthen agricultural value chains has been emphasized. The evolution of the approach to ANR, including the nature of the projects supported, is illustrated in Figure 9. Appendix 2 contains a list of all project loans and grants provided by ADB over the previous 20 years.

Pre-2010 Country Partnership Strategies

In the early 2000s when agriculture made a more substantial contribution to the economy as well as employment, the sector was a significant focus of ADB's interventions and largely aimed at poverty alleviation. Many projects were focused on livelihoods and worked directly with beneficiaries. This resulted in community-driven development-type projects addressing diversification, access to credit, livestock development, water and irrigation management, and local infrastructure needs. The strategies gradually increased focus on larger infrastructure related to rural roads, irrigation, and flood protection in line with government priorities. ADB also recognized the need to move beyond the farm level to support postharvest activities, marketing, and agribusiness development. In the latter phase, water-related and rural road infrastructure projects proved more successful and were viewed as an area of ADB's comparative advantage.

Country Partnership Strategy, 2011–2015

The CPS, 2011–2015 sought to narrow the focus of ANR sector support. Specifically, it focused ANR assistance on rural infrastructure and water resources management, which were areas where ADB had a comparative advantage and proven track record. Interventions in these areas were to support the government priorities of rural poverty reduction, food security, and gender equality. With this concentration, the CPS excluded several subsectors that were previously supported. The CPS stated, "ADB will reposition its public sector support away from a number of areas, such as participatory enhancement of livestock, crop diversification, small and medium-sized enterprises, and agribusiness, as well as civil service management and judicial reform."

Figure 9: Evolution of ADB's Bangladesh Country Partnership Strategies

Pre-2010 CPSs	CPS 2011–2015	CPS 2016–2020
• **Focus**: *broad*-rural development, livelihoods, diversification, intensification, productivity, agribusiness, and farm-to-market linkages • **Supporting Projects:** community-driven development, water resources management, irrigation, roads, livestock, diversification, agribusiness, flood protection	• **Focus**: *limited*-rural infrastructure and water resources management, and recognized Private Sector Operations Department work • **Supporting Projects:** irrigation, flood, climate resilience, roads, private sector agribusiness	• **Focus**: broadening livelihoods, productivity and profitability, diversification and higher-value crops, food safety and quality, and climate change **Supporting Projects:** • roads, rural infrastructure, private sector agribusiness

ADB = Asian Development Bank, CPS = country partnership strategy.
Source: Asian Development Bank.

Country Partnership Strategy, 2016–2020

The CPS, 2016–2020 continued support for rural infrastructure and water resources management and gave more prominence to agriculture and rural livelihoods. Its priorities were

(i)   raising productivity and profitability of agriculture,

(ii)   increasing diversification of crop and non-crop production for higher value and yield,

(iii)   maintaining food safety and quality, and

(iv)   tackling climate change impacts and improving disaster resilience.

The CPS also noted that increasing private sector participation in agro-processing value chains was important to accelerate agricultural diversification and improve market connectivity. It highlighted the importance of upgrading rural roads, improving maintenance, and developing agribusiness infrastructure. Resilience was to be supported through climate-proofing for more resilient infrastructure and integrated water resources management for greater food and water security. DRM was included as a crosscutting theme. This CPS returned to broader and more holistic support. This included a greater focus on a range of agriculture-related subsectors and an explicit focus on value chains. It also more explicitly recognized Bangladesh's vulnerability to climate change and related disasters, placing an emphasis on resilience.

Past Portfolio and Lessons

In terms of the corresponding portfolio, its focus generally followed the contracting and then expanding priorities of the CPSs. The country operations business plan for Bangladesh, 2021–2023 highlights the expanding priorities and a greater shift toward agriculture and support to various links in the value chain.

ADB-supported projects have generally performed well and offer several lessons. The Coastal Climate-Resilient Infrastructure Project, supported by the Pilot Program for Climate Resilience of the Strategic Climate Funds established under the Multi-Donor Climate Investment Fund, was completed in June 2019.[74] The project helped Bangladesh shift to a climate-resilient development path. The Chittagong Hill Tracts Rural Development Project[75] and a follow-up project[76] assisted rural infrastructure development and agribusiness development in one of the poorest and most geographically challenged regions of the country, drawing on community plans and priorities, and building local institutions involved in rural development. HVC production, value chain linkage, and agribusiness development have been supported under a series of crop diversification projects and agribusiness development projects. ADB also piloted weather index-based crop insurance as an innovative risk adaptation tool for farmers, including small farm households and received encouraging results.[77] All projects assessed under the latest country assistance program evaluation were expected to contribute to better gender equity. Benefits accruing to women have been substantial, especially from rural infrastructure projects, which have led to better access to employment, education, and health services as well as employment generated by projects for unskilled rural women both in the construction and maintenance phases of the projects.[78]

In water resources management, ADB assistance was provided in line with the National Water Policy and the National Water Management Plan, which adopted the basic principles of integrated water resources management, sustainable service delivery, and user participation in its water resources management support. ADB assistance covered a range of projects, including improvement of irrigation and water management facilities, post-disaster rehabilitation, and flood and riverbank erosion protection. ADB interventions were highly relevant, which complemented key elements of ADB's Water Policy including (i) institutional strengthening and improved information management; (ii) fostering integrated management of water resources; (iii) water conservation and improving system efficiencies; (iv) socially inclusive development principles to promote stakeholder consultation and participation at all levels; and (v) improved governance by promoting decentralization, building capacity, private sector participation, and strengthening monitoring and evaluation. ADB assistance supported the National Adaptation Program of Action and the Bangladesh Climate Change Strategy and Action Plan, which emphasized flood control and erosion risk reduction.[79] Institutional development and implementation of reform measures for efficient services delivery appeared to be slow.

The focus of recent water management projects has been innovation and sustainability. Activities under the Flood and Riverbank Erosion Risk Management Investment Program[80] reduced to some extent flood and riverbank erosion risks through innovative riverbank protection technologies that were developed under

---

[74]   ADB. Bangladesh: Coastal Climate-Resilient Infrastructure Project.

[75]   ADB. Bangladesh: Chittagong Hill Tracts Rural Development Project.

[76]   ADB. Bangladesh: Second Chittagong Hill Tracts Rural Development Project.

[77]   ADB. Bangladesh. Pilot Project on Weather Index-Based Crop Insurance.

[78]   Independent Evaluation Department. 2021. *Country Assistance Program Evaluation for Bangladesh*. Manila: ADB.

[79]   ADB. Bangladesh: Supporting Implementation of Bangladesh Climate Change Strategy and Action Plan (Subproject 2).

[80]   ADB. Bangladesh: Flood and Riverbank Erosion Risk Management Investment Program.

the ADB-funded predecessor project to other geographic areas.[81] The project contributed to strengthening institutional flood and riverbank erosion management systems for strategic management of the main rivers and for sustainable management of infrastructure. ADB has supported large- and small-scale irrigation schemes in Bangladesh through projects like the ongoing Irrigation Management Improvement Project, which addresses the lack of sustainable management and O&M by transferring them to qualified private operators and introducing innovative infrastructure modernization.[82] Another example is the Participatory Small-Scale Water Resources Sector Project, completed in 2018 and rated *successful* by ADB's Independent Evaluation Department.[83] Past projects like the Coastal Climate-Resilient Infrastructure Project have combined the specific infrastructure needs of coastal areas by improving road connectivity through road climate-proofing and construction of bridges and culverts, improving community markets, and enhancing climate change adaptation, not only through building capacity in communities but also by developing and improving cyclone shelters.[84]

For over 2 decades, ADB has focused on promoting agricultural diversification away from rice production.[85] This shift proved beneficial for small farmers but was not significant enough to move farmers toward commercialization and more formally into value chains. To further support diversification and better integrate farmers into value chains, these projects identified the following lessons to be addressed in future projects:

(i)   The integration of small farmers into value chains requires increased coordination between the multiple implementation partners and more substantial technical capacity of executing and implementing agencies.

(ii)   Specific interventions in market development and access to markets are needed to integrate farmer and producer groups into value chains.

(iii)   Greater private sector engagement is needed in postharvest, processing, cold chain, and logistics.

(iv)   It may be necessary to address specific policy and institutional barriers related to food safety, and research and extension, which may limit the effectiveness of project activities and leveraging of private sector resources in the value chain.

Although sovereign support to agricultural value chains was limited, nonsovereign support to the Programme for Rural Advancement Nationally (PRAN), an integrated food and agribusiness company, was successful.[86] The project was ADB's first nonsovereign agribusiness project since 1985 and primarily supported a starch and liquid glucose factory for use in processed foods. To source its starch, PRAN contracted cassava farmers, providing training, inputs, and a market. This was followed by a second loan to Sylvan Agriculture for processing of potato chips, potato flakes, and pasta through the financing of new food processing facilities.[87] ADB and PRAN have continued their partnership with a dairy project procuring raw milk from smallholder dairy farmers that aimed to support PRAN Dairy Limited following the disruptions in the food and agricultural supply chain that resulted from the COVID-19 crisis.[88]

[81]   ADB. 2002. Report and Recommendation of the President to the Board of Directors: Proposed Loan to Bangladesh for the Jamuna–Meghna River Erosion Mitigation Project. Manila.

[82]   ADB. Bangladesh: Irrigation Management Improvement Project.

[83]   Independent Evaluation Department. 2021. Validation Report: Participatory Small-Scale Water Resources Sector Project in Bangladesh. Manila: ADB.

[84]   ADB. Bangladesh: Coastal Climate-Resilient Infrastructure Project.

[85]   ADB. 2011. Completion Report: Northwest Crop Diversification Project in Bangladesh. Manila; and ADB. 2021. Completion Report: Second Crop Diversification Project in Bangladesh. Manila.

[86]   ADB. 2012. Report and Recommendation of the President to the Board of Directors: Proposed Loan Sylvan Agriculture Limited PRAN Agribusiness Project in Bangladesh. Manila.

[87]   ADB. 2018. Report and Recommendation of the President to the Board of Directors: Second PRAN Agribusiness Project in Bangladesh. Manila.

[88]   ADB. 2020. FAST Report: Emergency Working Capital Support to Dairy Value Chain Project in Bangladesh. Manila.

The ongoing rural connectivity and water management projects have shown progress in realizing key outcomes and outputs. Climate-resilient development has been mainstreamed in rural infrastructure. A rural road maintenance policy has been developed. Through water management cooperative associations, a fund for O&M for small-scale water resources management schemes has been created, and members contribute to it. This is being replicated in larger schemes. Sustainability of small-and medium-scale water resources management systems has been supported through the introduction of participatory irrigation management. Water management groups have been organized and their capacity strengthened. Management and O&M of the systems are being delegated to these groups.

Engagement of development partners in the agriculture and natural resources sector. In the ANR sector, active major multilateral development partners are ADB, International Fund for Agricultural Development, Islamic Development Bank, and World Bank Group. Bilateral development partners are Japan International Cooperation Agency, United States Agency for International Development, The Embassy of the Kingdom of the Netherlands, and KfW. Among the UN organizations, FAO and World Food Programme provide technical assistance. Since December 2017, ADB's ANR financing has declined to third, behind the World Bank and Japan, in terms of donor support to the sector.[89]

Local Consultative Group on Agriculture, Food Security, and Rural Development. Bangladesh has created several local consultative groups (LCGs) comprising representatives from bilateral and multilateral donors, NGOs, and research centers.[90] The LCGs are a platform for development partners to coordinate activities and harmonize and enhance aid effectiveness. The LCG on Agriculture, Food Security, and Rural Development (AFSRD) supports the government's agriculture sector transformation plan, which will incorporate issues of food safety, agro-processing, climate-smart agriculture, diversification, mechanization, and gender.[91]

Despite having some experience with agricultural development and food security, ADB's expertise largely lies in built infrastructure, traditionally associated with water resources management and rural development. Therefore, in support of the LCG-AFSRD, ADB aims to focus on these areas, while expanding its work on agricultural value chains and adaptation to climate change. It will continue working in coordination and partnership with other donors to foster synergies for greater development effectiveness in areas such as food safety, security, and nutrition. A summary of some of these institutions' priorities and strategic plans is presented in Box 2. Appendix 3 includes a list of recent and ongoing projects from the development partners.

---

[89]   Economics Relations Division.

[90]   The LCG-AFSRD is chaired by the Ministry of Agriculture and Ministry of Food and has as members development partners such as Food and Agriculture Organization of the United Nations, AsianFAO, World Bank, Consultative Group on International Agriculture Research centers (International Maize and Wheat Improvement Center, International Food Policy Research Institute, International Rice Research Institute, and WorldFish Research Institute), Bangladesh Bank, Building Resources Across Communities-Bangladesh, International Fund for Agricultural Development, International Union for Conservation of Nature, International Fertilizer Development Center, Bangladesh Agricultural Research Council, Switzerland Embassy, World Food Programme, Islamic Development Bank, and European Union.

[91]   The chairing of LCG-AFSRD is done by the Ministry of Agriculture, and a co-chair is assigned on a rotation basis. At time of writing in November 2022, the co-chair is FAO.

Box 2: Summary of Bangladesh's Development Partner Priorities and Strategic Plans

The World Bank promotes and provides financing to the agriculture sector in Bangladesh by taking a farm to fork approach, supporting projects on climate-smart agriculture, water management, dairy, and food storage facilities. The bank is moving from a project to program approach and aims to further support policy changes and institutional strengthening in the agriculture sector. It plans to develop an action plan for the support of the Bangladesh Delta Plan 2100 Investment Plan and has also partnered with the International Water Management Institute and private sector companies to strengthen the value chain of crops such as peanuts. Along with the Asian Infrastructure Investment Bank, it is also financing rural development via water, sanitation, and hygiene projects in rural areas. The World Bank is also supporting the Program on Agricultural and Rural Transformation for Nutrition, Employment, and Resilience in Bangladesh. This program's objective is to enable resilient production and marketing of high-value, safe, and nutritious food in Bangladesh.

The Food and Agriculture Organization of the United Nations has as its priorities the strengthening of urban food systems, promotion of digital villages, increasing the resilience of riverine systems, and supporting climate-smart agriculture. It is also supporting increased access to clean water and soil conservation in the Chittagong Hill Tracts and *haor* areas, and in partnership with the World Bank is working toward improvement of nutrition and smallholder competitiveness. The organization recognizes the importance of women as agricultural workers and is assessing how to maximize their participation.

The International Fund for Agricultural Development aims to (i) enable poor people in vulnerable areas to better adapt their livelihoods to climate change; (ii) help small producers and entrepreneurs benefit from improved value chains and greater market access; and (iii) empower marginalized groups, including poor rural women, both economically and socially. The fund places a strong emphasis on bringing youth to the agriculture sector and increasing interventions on *chars*.[a] While partnering with the Asian Development Bank as cofinancier for the development of climate-resilient small-scale water resources programs in several of Bangladesh's climate hot spots, the fund also supports the development of agricultural technology. With a loan of $66.5 million in 2018, the fund approved the Smallholder Agricultural Competitiveness Project, which will continue until 2024. The project aims to increase incomes and food and nutrition security for 250,000 rural households in southern Bangladesh. The fund's recent implementation experience highlights a need to include activities to enhance capacity of the implementing agencies as part of any project.

Islamic Development Bank was preparing a new country strategy in 2021, with its focus expected to be on food security, followed by infrastructure development, agricultural mechanization, and global value chains. Agriculture is the largest sector supported by the bank in its portfolio.

United States Agency for International Development is in the process of revising its own Indo-Pacific strategy, which will be focused on renewable energy, climate action, biodiversity, and wildlife preservation. In the past, its interventions focused on rice and crop diversification. Future projects are expected to include wetland conservation and hilsa fish conservation, and have a special focus on the Chittagong Hill Tracts.

The Netherlands is supporting flood protection infrastructure, such as the restoration of polders. It is phasing out interventions associated with food security and switching focus to value chain and climate-smart agriculture.

[a] A *char* is a tract of land surrounded by the waters of an ocean, sea, lake, or stream. In the dynamics of erosion and accretion in the rivers of Bangladesh, the sandbars emerging as islands within the river channel (island *chars*) or as attached land to the riverbanks (attached *chars*), often create new opportunities to establish settlements and pursue agricultural activities on them.

Source: Banglapedia, National Encyclopedia of Bangladesh.

III. ADB's Agriculture, Natural Resources, and Rural Development Sector Strategy

ADB's future ANR support in Bangladesh will respond to the evolving context and changes in the economy and build on its experience and lessons learned. The approach recognizes the ongoing structural transformation of the economy, changing role of the ANR sector, and the existential threat posed by climate change. The sector strategy, while aligning with the ADB CPS, will aim to further strengthen ADB ANR comparative advantages in supporting water resources management and rural infrastructure development and better leverage the knowledge and skills within the bank to support productivity improvements, value addition, and the commercialization of agriculture. It will also aim to expand support for ongoing and emerging needs identified in the previous CPS related to climate resilience and adaptation, as well as other crosscutting issues such as gender.

ADB's ANR sector strategy in Bangladesh will focus on promoting rural development and food security and improving rural livelihoods by supporting agricultural commercialization and increasing value addition by boosting the productivity and competitiveness of the sector. This strategy will be underpinned by four pillars: (i) agricultural commercialization and value chains, (ii) market connectivity, (iii) water and natural resources management, and (iv) crosscutting One ADB priorities. Environment and climate issues will be integrated throughout the program, given the importance of the natural resources base for agriculture. This strategy, through a One ADB approach, will also address other crosscutting concerns such as the policy and regulatory enabling environment for agribusiness, capacity, and gender. The ANR strategy is illustrated in Figure 10. Further details are provided in the results matrix in Appendix 4.

Figure 10: Agriculture, Natural Resources, and Rural Development 2021–2025 Strategy

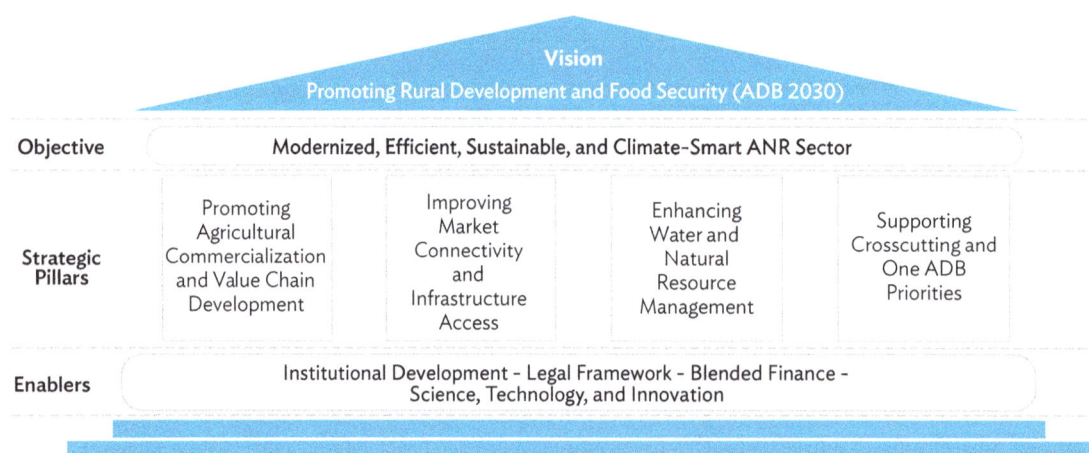

Vision
Promoting Rural Development and Food Security (ADB 2030)

Objective: Modernized, Efficient, Sustainable, and Climate-Smart ANR Sector

Strategic Pillars:
- Promoting Agricultural Commercialization and Value Chain Development
- Improving Market Connectivity and Infrastructure Access
- Enhancing Water and Natural Resource Management
- Supporting Crosscutting and One ADB Priorities

Enablers: Institutional Development - Legal Framework - Blended Finance - Science, Technology, and Innovation

ANR = agriculture, natural resources, and rural development.
Source: Asian Development Bank.

ADB's approach to ANR in Bangladesh is informed by the bank's Strategy 2030 (footnote 1) and its operational priority for rural development and food security.[92] The plan aims to boost rural development, agricultural value chains, and food security. ADB intends to support efforts improving market connectivity and agricultural value chain linkages. As with other developing member countries, ADB aims to increase agricultural productivity and food security by boosting farm and nonfarm incomes, introducing higher-value and stress-tolerant crop varieties, promoting the adoption of advanced technologies and climate-smart agricultural practices, and supporting the improvement of natural resources management standards. It also intends to help enhance food safety as an important requirement for commercialization and value addition within the sector.

The strategy aims to support the sector's role in Bangladesh's continued economic and structural transformation and shift toward upper-income status. This will entail aligning ADB policy, investment, technical, and capacity assistance with the development goals outlined in the Eighth Five-Year Plan and BDP2100. For agriculture, the focus is on modernizing to produce more nutritious, diversified, and safer food in a way that considers climatic and environmental challenges so as to improve adaptation, resilience, and sustainability. Given the country's unique set of environmental challenges and vulnerability to climate change, addressing these issues is a priority. To achieve these objectives, the strategy is supported by four pillars (areas of work) and a set of enablers (crosscutting and thematic issues).

The four pillars will support (i) aligning government policies, strategies, and incentives to modernize the ANR sector and address changing demands of food security; (ii) improving agricultural productivity, competitiveness, profitability, and uncertainty in rural areas; and (iii) promoting climate-smart agriculture and sustainable natural resources. Addressing these areas will contribute to 12 of the 17 SDGs.[93] They are also consistent with the ADB ANR sector objectives from the ADB results framework (Appendix 4).

Pillar 1: Promoting Agricultural Commercialization and Value Chain Development

In supporting agricultural commercialization and value chains, a main objective will be to increase smallholder farmer productivity and competitiveness for moving from subsistence/semi-subsistence to commercial agriculture, and meeting value chain requirements. Special attention will be given to women farmers to address their specific challenges. ADB will support productivity increases and greater sustainability by assisting smallholders to access new or improved inputs (e.g., crop varieties, irrigation systems, technologies, agricultural practices, and weather forecasting and market information) that are better adapted to evolving climatic and market conditions. ADB will increase support to smallholder farmers to meet market quality and volume requirements for HVCs and those with potential for processing and value addition. Interventions will identify and address key weaknesses along domestic and export value chains, particularly if these improvements will benefit or increase the involvement of smallholder farmers. Specific attention will be given to addressing changing domestic demands, producing better quality and more nutritious fresh and processed foods, and integrating food safety practices and standards. Value chain activities will also target specific challenges such as logistics, marketing support, and aggregation. For example, to address the challenges of aggregation and economies of scale, efforts will be made to strengthen connections between farmers and intermediaries (e.g., cooperatives) with value chain participants and access to finance. Wherever possible, farmer-producer organizations may be used as a key entry point for engaging farmers and linking them to value chain actors such as processors and retailers, which will enhance prices smallholder farmers receive. At the same time, ADB will continue to support private sector operations within the ANR sector, with an emphasis on value chains that are potentially inclusive and have an important development impact.

---

[92]   ADB. 2019. *Strategy 2030: Operational Plan for Priority 5. Promoting Rural Development and Food Security, 2019–2024.* Manila.

[93]   SDGs 1, 2, 3, 4, 5, 6, 7, 11, 13, 14, 15 and 17.

ADB may undertake strategic diagnostic assessments to identify specific commodities and upstream areas related to the enabling environment for support. For commodities, this would identify key opportunities and constraints as well as main public and private sector actors. ADB will also engage and consult with downstream private sector actors in the value chain (e.g., processors and retailers). By making connections with key firms, partnerships throughout the value chains can be strengthened. This will also reveal opportunities and constraints for value chain development, and the inclusion of smallholder farmers and producers' organizations. It may also identify larger firms for partnerships with sovereign operations and support from ADB's Private Sector Operations Department.

In the enabling environment, diagnostics may consider policy, regulatory, or institutional barriers (e.g., related to food safety) that inhibit development of the sector or specific value chains. A key feature of these assessments will be to consider the climate-related risks to key activities. Support will be given to modernize and strengthen agricultural institutions (e.g., research, education, and extension) and to improve sector resilience and its adaptation to changing climate and market conditions. This could include support for research and development (R&D), or the introduction and use of new technologies—particularly in developing climate-suitable varieties, climate-smart agriculture, and more resilient on-farm approaches.

Within ADB, there will be greater collaboration between sovereign and nonsovereign operations for financial intermediaries lending to agri-enterprises and providing other financial services, contract farming, and agribusiness development. Joint missions and diagnostics may be considered to identify opportunities supporting private sector development. ADB will also engage in dialogue with the government as policies, government priorities, and expenditures on agricultural inputs (e.g., extension) are important determinants of crop choice and productivity. Further government policies and practices are essential for establishing the policy and regulatory environment to support corridor, cluster, and value chain development (e.g., food safety and export certifications). ADB will also support food safety in partnership with the government and private sector. This could include introducing good practices and certification, particularly for high-value horticulture crops or those with export potential.

Pillar 2: Improving Market Connectivity and Infrastructure Access

For market connectivity and access, the focus will be on supporting market infrastructure for agricultural products and the development of value chains. A core part of this will be to build transportation infrastructure, improving market linkages and farmers' access to inputs and other value chain actors. This will include rehabilitating, upgrading, climate-proofing, and improving the maintenance of rural roads, which are essential for connecting rural communities to economic opportunities and social and administrative services. These investments will include climate-resilient designs with robust and effective quality control arrangements during construction, plus adaptive management and maintenance to increase resilience and improve sustainability.

In addition to rural roads, supporting inland water transport is another area of possible development. Inland water transport may be more efficient and practical for moving agricultural products to market: its costs and environmental impact can be lower than other modes of transport. ADB could consider fostering inland waterway development with all-season flows for conservation and navigation, as well as inland port O&M.

Along such improved roads and inland waterways, ADB may consider a corridor development approach, where additional infrastructure or technical assistance is provided to support agricultural growth in the area particularly related to HVCs and commodities that have a comparative advantage. In other zones, such as Chittagong Hill Tracts, an area development approach may be preferred.

ADB will target investments that improve rural connectivity. Market infrastructure and logistics services supporting agribusiness will be developed to reduce postharvest losses. Investments in market, storage, and processing infrastructure will ensure that produce is washed, graded, packed, pre-cooled, and transported via a coordinated logistics and distribution system to the consumer. Along the value chain from farm to market, ADB can consider investing in facilities for sorting, aggregating, and bulking commodities. High-quality modern storage facilities and markets are also needed along with improved food safety standards and certification. To further minimize postharvest losses and damage from frequent handling, postharvest management and the introduction of good practices can be supported. Basic drying, processing, and bagging or packaging can also be considered to increase shelf life.

ADB will encourage both public and private stakeholders in coordinating and sharing information for better management of market infrastructure and integration of producers into the value chains. ADB will also support institutional arrangements and capacity for managing and maintaining the infrastructure, particularly where there are opportunities to include the public and private sectors. This is needed to sustain improved market linkages and ensure their resilience to climatic shocks; it may include technical training, twinning arrangements, institutional reforms, and the introduction of new processes and information and communication technology (ICT).

Pillar 3: Enhancing Water and Natural Resources Management

Given the importance of natural resources for agriculture and its vulnerability to climate change impacts, integrated and strategic water, land, and natural resources management will be central to ADB's support. Throughout the country, each region will need solutions tailored for its own specific combination of agronomic and climatic challenges. Interventions in the irrigation and water resources sector will support adaptation to climate change, modernization of irrigation systems with increased water use efficiency, and sustainable management of water resources and related infrastructure. While addressing these challenges, interventions will be more inclusive with increased community participation, and improved irrigation service delivery, which will increase accountability to farmer organizations. The success of ADB's small-scale water resources projects shall be enhanced with a sharper focus on climate resiliency and decentralized, community-based water management.

Increased emphasis on a climate-resilient approach will also extend to upgrading support for flood control, riverbank protection and stabilization, and adaptive basin management. When feasible, priority will be given to the most vulnerable areas such as *char, haors*, wetlands, polders, and coastal areas. These investments and more responsive management are essential for improving resilience and supporting agricultural, inland fisheries, and other economic activities along the rivers. The mitigation of flood, coastal, and riverbank erosion will continue to be a priority, reinforced by adding new dimensions to natural resources management and nature-based solutions (NBS) such as those described below.

Climate change challenges require a more holistic approach, with deeper comprehension of water and natural resources management. ADB may consider diagnostics and risk mapping to better understand not only the vast river systems in Bangladesh and potential climate impacts, including maladaptation, but also to consider the new irrigation needs that will follow modernization, crop diversification, and changing cropping patterns. The sector will also benefit from research into the impact of agrochemical discharges on water quality, and the advantages of regenerative agriculture for ADB's future investment. These diagnostics could also include more integrated land-use planning, and tools such as strategic environmental assessments for flood risk management and early warning systems. This may also propagate NBS and other green infrastructure to complement traditional frameworks, with remote sensing technologies. For the environment, climate resilience, and watershed management, ADB may also consider conducting upstream diagnostic studies and making investments in soil regeneration, reforestation, wetlands and coastal restoration, community forests, fisheries, and further harnessing the blue economy potential. Such supportive activities will be executed through technical assistance or investment programs.

Institutional capacity strengthening, for strategic O&M planning and implementation, will remain a key priority. The lack of financially sustainable O&M for large-scale water resources management needs to be addressed, by transferring these functions to qualified private operators, developing viable business models, strengthening financial team capacity, and innovating infrastructure. For sustainable O&M of small- and medium-scale, participatory management and decentralized operations by water user groups will continue to be pursued. In natural resources management, the focus will be on improving disaster risk management (DRM) in terms of land use, planning, and early warning systems.

Pillar 4: Supporting One ADB Priorities

This pillar supports the One ADB priority of greater integration of its operations to achieve better development outcomes in line with Strategy 2030. It recognizes that ANR overlaps with the bank's other priority sectors. Key areas for potential collaboration and joint activities lie within these sectors:

(i)   energy (e.g., solar-pumped groundwater);

(ii)   education (e.g., tertiary education);

(iii)   finance (e.g., rural micro- and SME-lending);

(iv)   public sector management (e.g., safety nets and food security);

(v)   transport (e.g., reduced commodity transport costs through rail and inland waterways); and

(vi)   urban (e.g., integrated watershed management and urban–rural linkages).

Priorities are financial support, including more farmers, and identifying better financial services for agribusinesses in the value chain.

Recognizing agriculture's multisector nature and complexity, the assistance program will also improve institutional capacity and coordination among government agencies, development partners, beneficiaries, private sector organizations, and others. Expanding partnerships with the private sector will be vital, given the importance of value addition. Essentially a private sector activity, agriculture is ideal for greater collaboration with the Private Sector Operations Department in addressing development challenges, supporting joint diagnostics, and identifying areas for investment.

This pillar will also strengthen several crosscutting ADB priorities that are needed to achieve the overall CPS sector objectives. ANR sector interventions will empower the important drivers of change by improving (i) governance and capacity development, (ii) gender equity and mainstreaming, (iii) maternal health and nutrition, (iv) knowledge solutions, (v) innovative R&D to develop climate-resilient production technologies, and (vi) partnerships. As ANR is highly vulnerable to the effects of climate change, adaptation considerations will be mainstreamed in ADB support to help Bangladesh follow a path of more climate-resilient agricultural development. Specific attention will be given to supporting knowledge and sharing valuable lessons from successful investments with the potential for replication (e.g., flood and riverbank erosion protection).

Figure 11: Key Enablers for Implementing the Strategic Framework

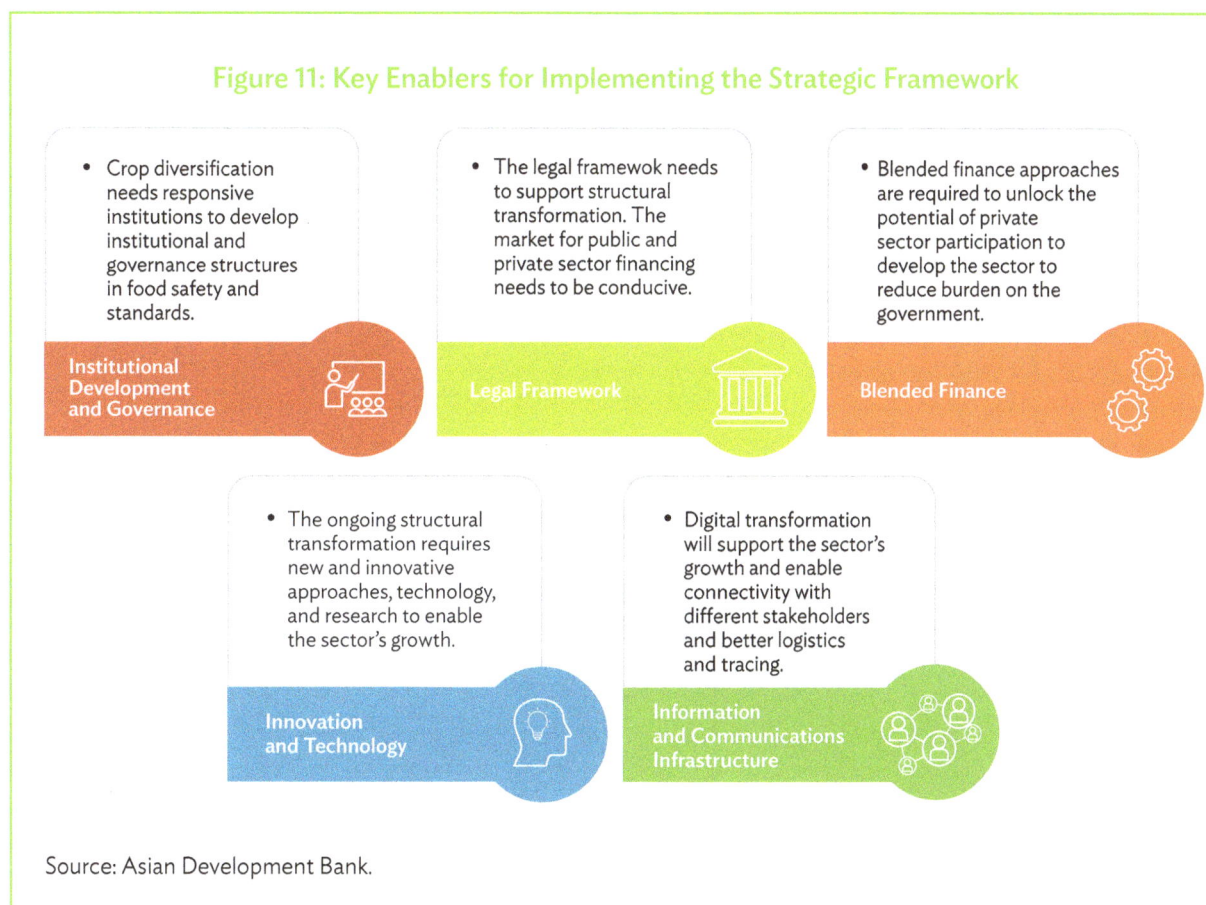

- Crop diversification needs responsive institutions to develop institutional and governance structures in food safety and standards.

Institutional Development and Governance

- The legal framewok needs to support structural transformation. The market for public and private sector financing needs to be conducive.

Legal Framework

- Blended finance approaches are required to unlock the potential of private sector participation to develop the sector to reduce burden on the government.

Blended Finance

- The ongoing structural transformation requires new and innovative approaches, technology, and research to enable the sector's growth.

Innovation and Technology

- Digital transformation will support the sector's growth and enable connectivity with different stakeholders and better logistics and tracing.

Information and Communications Infrastructure

Source: Asian Development Bank.

Guiding principles for implementing the strategic framework. Several principles will guide ADB in tackling evolving systemic challenges, to make greater development impacts in ANR—particularly in advancing food systems and value chains, improving the environment and ecosystems services, and ensuring stronger climate resilience. ADB will strive to manage more complex development challenges by using innovative approaches to deliver enhanced solutions. In pursuing greater innovation and value addition, ADB will manage risk appropriately, and recognize the importance of enablers in its programming (Figure 11).

In executing the strategy, ADB's approach will build on its areas of demonstrated success, comparative advantage, and the directions identified in its strategic pillars. Further subsector diagnostics will highlight specific targets for project and program investments. The process will involve a rigorous analysis of data, policies, and institutions. It will also focus closely on collaboration, coordination, and consultation with key partners and stakeholders in the sector. While engaging new partners selectively and incrementally, ADB will continue to work with experienced implementing agencies. The ANR program will also optimize diverse lending instruments, using investment loans for innovative areas to develop a proof of concept, while considering policy-based, results-based, and sector loans for replicating and upscaling successful initiatives. Finally, recognizing the increasing complexity of challenges facing ANR, ADB will also encourage greater collaboration among its technical specialists working in the private, urban, water, education, and finance sectors.

Appendixes

Appendix 1: Problem Tree for the Agriculture, Natural Resources, and Rural Development Sector

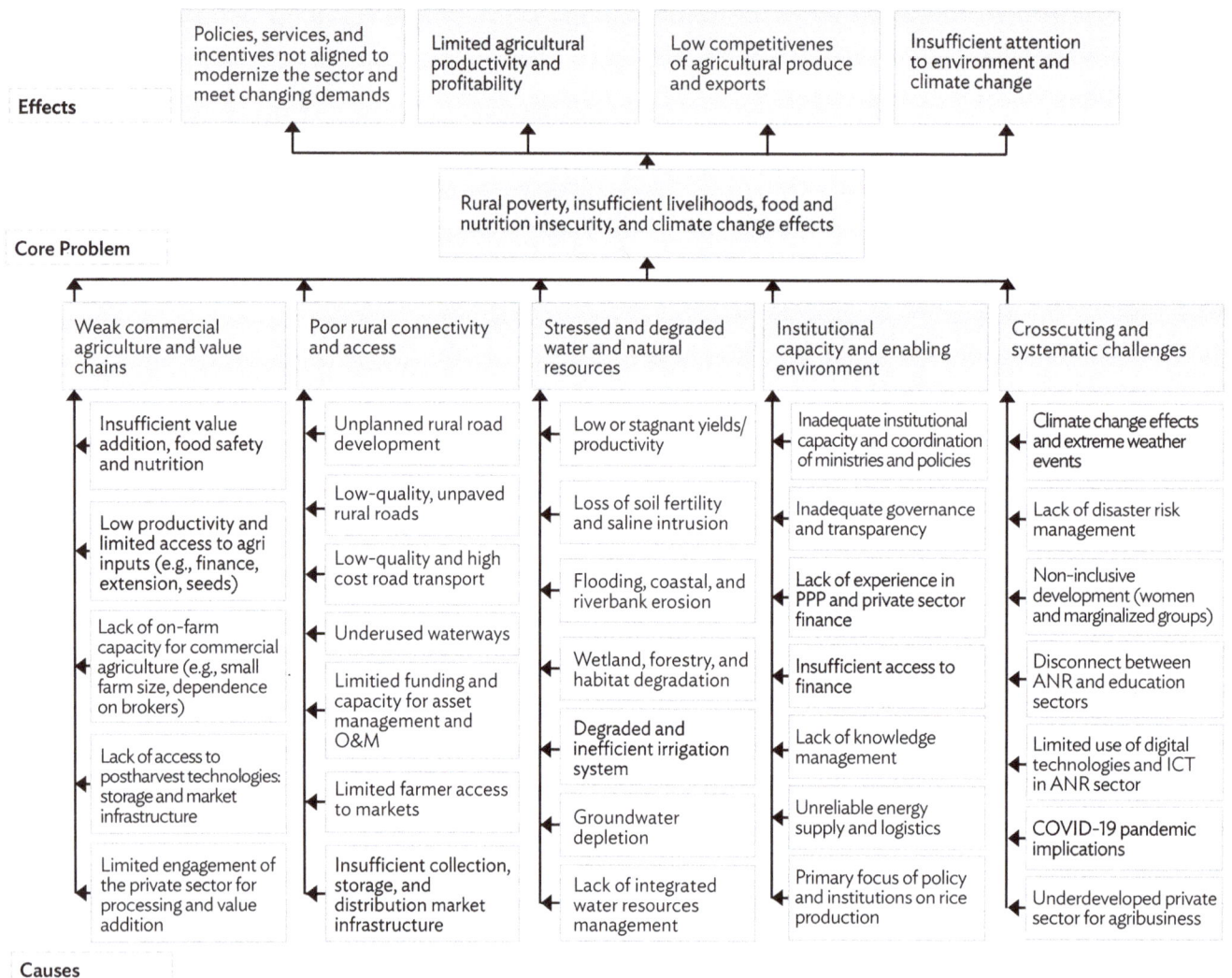

Effects

Policies, services, and incentives not aligned to modernize the sector and meet changing demands	Limited agricultural productivity and profitability	Low competitivenes of agricultural produce and exports	Insufficient attention to environment and climate change

Core Problem

Rural poverty, insufficient livelihoods, food and nutrition insecurity, and climate change effects

Weak commercial agriculture and value chains	Poor rural connectivity and access	Stressed and degraded water and natural resources	Institutional capacity and enabling environment	Crosscutting and systematic challenges
Insufficient value addition, food safety and nutrition	Unplanned rural road development	Low or stagnant yields/productivity	Inadequate institutional capacity and coordination of ministries and policies	Climate change effects and extreme weather events
Low productivity and limited access to agri inputs (e.g., finance, extension, seeds)	Low-quality, unpaved rural roads	Loss of soil fertility and saline intrusion	Inadequate governance and transparency	Lack of disaster risk management
Lack of on-farm capacity for commercial agriculture (e.g., small farm size, dependence on brokers)	Low-quality and high cost road transport	Flooding, coastal, and riverbank erosion	Lack of experience in PPP and private sector finance	Non-inclusive development (women and marginalized groups)
	Underused waterways	Wetland, forestry, and habitat degradation	Insufficient access to finance	Disconnect between ANR and education sectors
Lack of access to postharvest technologies: storage and market infrastructure	Limitied funding and capacity for asset management and O&M	Degraded and inefficient irrigation system	Lack of knowledge management	Limited use of digital technologies and ICT in ANR sector
	Limited farmer access to markets	Groundwater depletion	Unreliable energy supply and logistics	COVID-19 pandemic implications
Limited engagement of the private sector for processing and value addition	Insufficient collection, storage, and distribution market infrastructure	Lack of integrated water resources management	Primary focus of policy and institutions on rice production	Underdeveloped private sector for agribusiness

Causes

ANR = agriculture, natural resources and rural development; COVID-19 = coronavirus disease; ICT = information and communication technology; O&M = operation and maintenance; PPP = public–private partnership.
Source: Asian Development Bank.

Appendix 2: Sovereign and Nonsovereign Agriculture, Natural Resources, and Rural Development Sector Projects in Bangladesh, 2000–2021

No.	Project	Date of Approval	Amount ($ million)	Implementation Period
	SOVEREIGN OPERATIONS			
1771	Chittagong Hill Tracts Rural Development Project	26 Oct 2000	30.0	26 Oct 2000 to 22 Feb 2010
1782	Northwest Crop Diversification Project	21 Nov 2000	45.0	21 Nov 2000 to 12 Jan 2010
1941	Jamuna-Meghna Riverbank Erosion Mitigation Project	25 Nov 2002	42.17	25 Nov 2002 to 30 Jun 2011
2070	Second Participatory Livestock Development Project	19 Dec 2003	20.0	19 Dec 2010 to 30 Jun 2012
2190	Agribusiness Project	27 Oct 2005	42.5	27 Oct 2005 to 28 Nov 2012
2200	Southwest Area Integrated Water Resources Planning and Management	23 Nov 2005	20.0	23 Nov 2005 to 31 Dec-2015
2254	Second Rural Infrastructure Improvement Project	18 Aug 2006	96.10	18 Aug 2006 to 26 Dec 2013
2430	Emergency Assistance for Food Security	22 Jul 2008	170.0	22 Jul 2008 to 21 Jun 2010
2542/ 8248/ 8250	Participatory Small-Scale Water Resources Project	04 Sep 2009	55.0	04 Sep 2009 to 19 Nov 2020
2696	Sustainable Rural Infrastructure Improvement Project	11 Nov 2010	60.0	11 Nov 2010 to 24 Sep 2018
2649	Second Crop Diversification Project	30 Jun 2010	40.0	30 Jun 2010 to 23 Jul 2018
2763	Second Chittagong Hill Tracts Project	14 Jul 2011	55.0	14 Jul 2011 to 30 Jun 2021
2913	Coastal Climate Resilient Infrastructure	28 Sep 2012	20.0	28 Sep 2012 to 30 Jun 2019
9172	Pilot Project on Weather Index-based Crop Insurance	27 Mar 2013	2.0	27 Mar 2013 to 11 Oct 2018
3138	Flood and Riverbank Erosion Risk Management Program (ongoing)	03 Jul 2014	65.0	03 Jul 2014 to 31 Mar 2021
3135	Irrigation Management Improvement Project (ongoing)	03 Jul 2014	46.0	30 Jun 2014 to 30 Jun 2022
3302	Southwest Area Integrated Water Resources Planning and Management – Additional Financing	30 Sep 2015	45.0	30 Sep 2015 to 31 Dec 2022
3731/ 3732/ 3932	Rural Connectivity Improvement Project (ongoing)	05 Nov 2018	300.0	5 Nov 2018 to 31 May 2024
4107/ 0799	Flood and Riverbank Erosion Risk Management Investment Program – Project 2 (ongoing)	06 Sep 2021	174.89	06 Sep 2021 to 26 Jun 2024
4120	Irrigation Management Improvement Project-Additional Financing (ongoing)	30-Sep-2021	13.5	30-Sep-2021 to 31-Dec-2023

continued on next page

Appendix 2   *continued*

Project	Date of Approval	Amount ($ million)	Implementation Period
NONSOVEREIGN OPERATIONS			
BAN: Programme for Rural Advancement Nationally Agribusiness Project	08 Aug 12	25.1	14 Nov 2012 to 27 Jan 2020
BAN: Second Programme for Rural Advancement Nationally Agribusiness Project	12 Dec 18	14.2	19 Mar 2019 to 15 Dec 2025
BAN: Emergency Working Capital Support to Dairy Value Chain Project	15 Jul 20	10.0	14 Sep 2020 to 15 Mar 2023
REG: New Hope COVID-19 Working Capital Support Project	08 Dec 20	2.53	(3-year tenor)

BAN = Bangladesh, COVID-19 = coronavirus disease, REG = regional.
Source: Asian Development Bank.

Appendix 3: Mapping of Development Partners

Other Development Partners' Strategies and/or Main Activities (ongoing projects/programs and approval year)			
Multilateral Institutions and the United Nations System		**Bilateral Institutions**	
AIIB	• Rural Water, Sanitation and Hygiene for Human Capital Development Project (2020)	Danida	• Thematic Agriculture Growth and Employment Programme (2016)
FAO	• Technical Assistance to Smallholder Agricultural Competitiveness Project (2019) • Integrated Agricultural Development for Nutrition Improvement in the NW Region (2019) • Community-Based Climate-Resilient Fisheries and Aquaculture Development in Bangladesh (2019)	KOICA	• Improving living conditions of rural and urban poor through climate-adaptive and affordable housing technologies and WASH interventions (2018)
IFAD	• National Agricultural Technology Programme – Phase II Project (2015) • Promoting Resilience of Vulnerable Through Access to Infrastructure, Improved Skills and Information (2017) • Smallholder Agricultural Competitiveness Project (2018) • Rural Microenterprise Transformation Project (2019) • Climate-Resilient Small-Scale Water Resources Management Programme (2022, cofinanced with ADB)	EU	• Pro-Poor Growth of Rural Enterprises through Sustainable Skills Development (2016)

continued on next page

Appendix 3   *continued*

Other Development Partners' Strategies and/or Main Activities (ongoing projects/programs and approval year)			
Multilateral Institutions and the United Nations System		Bilateral Institutions	
IsDB	• Rangpur Division Agriculture and Rural Development Project (2016) • Rural Access Road Improvement in Sylhet Division project (2014) • Agricultural Support for Smallholders in South-Western Region of Bangladesh Project (2013)	Switzerland	• Agricultural and Disaster Insurance Programme
CGIAR ICRISAT IWMI	• Solar Irrigation for Agricultural Resilience (2019) • Flood Index Insurance – Bangladesh (2020) • Modelling ANR in Water (2019)	The Netherlands	• Water Support and Water Partnership (2017) • Water Management Knowledge & Innovation Program (2017) • Scaling Up Rice Fortification (2017)
IFC	• Financing of additional capital expenditures for wafer and aseptic lines, as well as maintenance capital expenditures at Programme for Rural Advancement Nationally entity Mymensingh Agro Limited (2020) • Capacity expansion of the PRAN Group at two sites in Natore and Rajshahi (2018)	JICA	• Food Value Chain Improvement Project (2020) • Small Scale Water Resources Development Project (2017)
MIGA	• Guarantee for Ghorasal Polash Urea Fertilizer Project	USAID	• Rice and Diversified Crops (2016)
World Bank	• Program on Agricultural and Rural Transformation for Nutrition, Employment, and Resilience in Bangladesh (PARTNER) (TBC – 2023) • Climate-Smart Agriculture and Water Management (2021) • Rural Water, Sanitation and Hygiene for Human Capital Development Project (2020) • Additional Financing for Modern Food Storage Facilities Project (2020) • Livestock and Dairy Development Project (2018) • Bangladesh Weather and Climate Services Regional Project (2017–22)	KfW	• Climate Resilient Infrastructure Mainstreaming (2018)
WFP	• Technical Assistance and Advocacy for Scaling-Up Post-Harvest Rice Fortification (2017)		

Note: This list is based on initial but limited consultations and some donors and projects may not be included.

ADB = Asian Development Bank; AIIB = Asian Infrastructure Investment Bank; CGIAR = Consultative Group on International Agricultural Research; EU = European Union; FAO = Food and Agriculture Organization of the United Nations; ICRISAT = International Crops Research Institute for the Semi-Arid Tropics; IFAD = International Fund for Agricultural Development; IFC = International Finance Corporation; IsDB = Islamic Development Bank; IWMI = International Water Management Institute; JICA = Japan International Cooperation Agency; KOICA = Korea International Cooperation Agency; MIGA = Multilateral Investment Guarantee Agency; TBC = to be confirmed; USAID = United States Agency for International Development; WASH = Water, Sanitation, and Health; WB = World Bank; WFP = World Food Programme.

Source: Asian Development Bank.

Appendix 4: Results Framework

	Pillar 1: Promoting Agricultural Commercialization and Value Chain Development	Pillar 2: Improving Market Connectivity and Access	Pillar 3: Water and Natural Resources Management	Pillar 4: One ADB Priorities
Outcome	More inclusive value chains	Improved market access	Increased productivity, resilience, and sustainability	Increased cross-sector integration for improved productivity and livelihoods
Indicators	2.3 Women represented in decision-making structures and processes (number) 5.2.3 Agribusinesses integrating farmers in efficient value chains (number) 5.1.4 Rural economic hubs supported (number) Commercial farming land supported (hectares) 5.3.3 Modern knowledge-intensive corporate farming models introduced (number) 5.2.4 Food safety and traceability standards improved (number)	5.2 Farmers with improved market access (number) 5.2.1 Wholesale markets established or improved (number) 5.2.2 Storage, agri-logistics, and modern retail assets established or improved (number) 22 Paved roads (kilometers per 10,000 people)	2.5 Women and girls with increased resilience to climate change, disasters, and other external shocks (number) 5.3 Land with higher productivity (hectares) 3.2 People with strengthened climate and disaster resilience (number) 5.3.1 Land improved through climate-resilient irrigation infrastructure and water delivery services (hectares) 5.3.2 Farmers using quality farm inputs and sustainable mechanization (number) 3.3.4 Solutions to conserve, restore, and/or enhance terrestrial, coastal, and marine areas implemented (number) 3.3.3 Terrestrial, coastal, and marine areas conserved, restored, and/or enhanced (hectares) 3.3.5 Sustainable water–food–energy security nexus solutions implemented (number)	11 Committed operations that support climate change mitigation and adaptation (%) (sovereign and nonsovereign) 12 Financing for climate change mitigation and adaptation ($ billion, cumulative) (sovereign and nonsovereign) 5.1 People benefiting from increased rural investment or improved rural infrastructure (number) 20 Proportion of population using basic drinking water and sanitation services (%) – rural 2.1.3 Women-owned or -led SME loan accounts opened or women-owned or -led SME end borrowers reached (number) 1.1.1 People enrolled in improved education and/or training (number)

SMEs = small-and medium-sized enterprises.

Notes: Two-digit references relate to selected Level 2A results framework indicators. The first number refers to the operational priority in Strategy 2030 (https://www.adb.org/documents/results-framework-indicator-definitions). Three-digit references relate to the associated tracking indicators for Level 2A (https://www.adb.org/documents/tracking-indicator-definitions).

Source: Asian Development Bank.

www.ingramcontent.com/pod-product-compliance
Lightning Source LLC
Chambersburg PA
CBHW042035220326

41599CB00045BA/7401